Electrical
Wiring:
Residential

Electrical Wiring: Residential

William J. Whitney
Albert Lea Area Vocational Technical Institute
Albert Lea, Minnesota

John Wiley & Sons

New York
Chichester
Brisbane
Toronto

Library of Congress Cataloging in Publication Data:

Whitney, William J., 1918-
Electrical Wiring-residential.

Includes index.
1. Electrical wiring, Interior. I. Title.
TK3285.W48   621.319′24   78-12869
ISBN 0-471-03578-5

Printed in the United States of America

10 9 8 7 6 5 4 3 2 1

To my wife
Elizabeth
and my two sons
Bill Jr. and Charles

Preface

As a vocational teacher, I am aware of the need for student materials that explain the nonmathematical theory behind electrical devices and equipment. Drawing on my experience as a journeyman electrician, electrical contractor, apprentice, adult educator, and a Vocational Technical Institute instructor, I have aimed, in this book, to explain as simply as possible electrical wiring applied to a single-family dwelling. I have tried to downplay the use of complex technical terms and to emphasize the reasons for electrical installation practices. Much of the text and many of the illustrations and examples have been reviewed by manufacturers and others to insure that the content is up to date and corresponds to the 1978 National Electrical Code.

It is my hope that high school students, vocational students, Vocational Technical Institute students, apprentices, tradespeople, and adult classes will find this book a useful tool for understanding residential wiring.

I think you will enjoy the book's format. It includes instructional objectives and several questions at the beginning of the chapter, designed to focus your attention on what's important, point you in the right direction, and make sure you understand the material in each chapter.

As you read the material, you will notice that environmental and energy problems are considered.

I gratefully acknowledge the support of the Wiley staff and the tireless work of Ms. Irene Holecek who typed the manuscript and its many revisions. I am deeply grateful to Mrs. Barbara Vogelsang for her editorial help; she read the manuscript of this book with meticulous care.

William J. Whitney

Acknowledgments

Many manufacturers of equipment have contributed illustrations, photos, written materials, and suggestions and their efforts, time, and assistance have been of great value.

I would like to thank each of them for their contribution.

City of Albert Lea, Minnesota

State of Minnesota

B and K Lumber Company.   Northwood, Iowa

Underwriters Laboratories, Inc.   Chicago, Illinois

Raco, Inc.   South Bend, Indiana

Lennox Industries, Inc.   Marshalltown, Iowa

Brown and Sharpe Mfg. Co.   Providence, Rhode Island

Martin Industries.   Florence, Alabama

Ideal Industries.   Sycamore, Illinois

General Electric Company.   Plainville, Connecticut

Pass and Seymour.   Syracuse, New York

The Singer Company.   Carteret, New Jersey

Blackhawk Industries.   Dubuque, Iowa

Midland-Ross Corporation.   Pittsburgh, Pennsylvania

Square D. Company.   Lexington, Kentucky

Nutone Division, Scovill Mfg. Co.   Cincinnati, Ohio

Cutler-Hammer, Inc.   Milwaukee, Wisconsin

Essex International.   Fort Wayne, Indiana

Albert Lea Vocational Technical Institute.   Albert Lea, Minnesota

Elektra Systems, Inc.   Farmingdale, New York

Contents

Electrical
Wiring:
Residential

Part 1 Introduction

Chapter 1 **General Information**

Instructional The purpose of the book is to explain the how and why of the elec-
Objectives trical installation so that it will be safe, economical, and convenient
to the builder.

1. To learn how the electrical wiring information is conveyed to the electrician.
2. To develop familiarity with working drawings, specifications, electrical symbols, and drawing notations used in architectural plans.
3. To become aware of electrical standards.
4. To be aware of the need of building permits and inspections.
5. To learn about listing and labeling of products.

Self-Evaluation Test your prior knowledge of the information in this chapter by
Questions answering the following questions. As you read the chapter, watch
for the answers. When you have completed the chapter, return to
this section and answer the questions again.

1. What is a working drawing?
2. How does the architect convey instructions to the electrician on a job site?
3. Why is cooperation between the crafts essential?
4. What name is given to the reproduction of the working drawing?
5. How would you define a floor plan?
6. How does the floor plan differ from a plot plan?
7. Why is it important for the electrician to understand the specifications before starting the job?
8. What is the purpose of using standard electrical symbols?
9. Why do we need electrical standards to protect the builder?
10. What is the National Electrical Code?

**1-1
The Working Drawing** Generally a prospective owner or builder will meet with an architect to discuss the planning of the new house. The architect will ask such important questions as:

1. How large a family do you have?
2. What is the size and shape of your lot?
3. Where is your lot located?
4. What quality materials do you want used?
5. What name brand of equipment do you prefer?
6. Do you intend to select and purchase the fixtures?
7. In what price range do you consider building?
8. What is your preference in layout and design?

When these questions are answered by the owner or builder, the architect is ready to create a set of drawings containing all of the information and dimensions necessary to successfully complete the project. These drawings are referred to as working drawings. Reproduction of the drawings are called a set of blueprints. The architect uses this set of blueprints to convey instructions to all of the crafts who are to plan, erect, and complete the structure.

The newcomer, apprentice, or electrician finds it an advantage to be able to read a set of blueprints early in the job experience. When this ability is acquired, that person becomes a part of a team representing many skills working out a series of construction problems. "Each must know now to "take off" dimensions accurately so that all of the outlets and equipment are located according to the blueprint."

All crafts must follow the print carefully if the work is to progress smoothly. Cooperation between the workers is essential.

**A Set of Working
Drawings** Figure 1.1 illustrates the working drawing, which consists of:

1. *Floor Plan:* This plan shows the layout of rooms, partitions, windows, doors, cabinets and bath fixtures. The architect usually draws a floor plan that includes all the electrical outlets, equipment, devices, and circuits.
2. *Elevation Drawing:* Shows what the outside of the house will look like.
3. *Plot Plan:* Tells where the house is located on the lot and gives information about the grade level and utilities such as gas, water, and electric service.
4. *Detail Drawing:* Is usually drawn to a larger scale than the other

drawings in order to show special features.

5. *Sectional Drawing:* Taken vertically through an exterior wall; may be drawn on the same sheet as one of the elevation drawings.

With the use of a working drawing the electrician has several guidelines to follow that actually tell how to proceed with the work of providing an efficient electrical installation.

1-2 Specifications

Specifications are a vital part of the building plans. They are the rules governing the type, kinds, quality, and colors of the materials used and the work to be performed by all of the crafts on the construction project. When the details are clear in the specifications, there is little room for misunderstanding or for misinterpretation of the plans. The specifications enable the electrical contractor to estimate the bid in terms of the legal responsibilities, guarantees, permits, inspections, brand names, and quality of the materials required. Lighting fixtures are usually chosen by the owner. However, they are usually purchased through the contractor, so the architect specifies a certain amount of money in the specs for a fixture allowance.

Specifications for a job are written in sections. Each section pertains to a different craft.

It is most important for the electrician to review all of the drawings and thoroughly understand the specifications before starting the job.

1-3 Symbols and Notations

It is essential that the architect and the craftsmen on the job understand each other. The architect will place all of the information possible on the drawing; the electrician must know how to interpret the drawing and apply it to the construction project.

The architect is working at a small scale and must omit many lines so that the blueprint may be readable. In order to do this, standard symbols are used to represent locations and types of materials to be used in the building. Most symbols and notations have a standard interpretation in the United States.

The notation that the architect places on the drawing next to a specific symbol provides information on the type, size, or quality of the device required. It is important that you identify symbols and understand the notations regarding the interpretation of all electrical symbols on the blueprint (Figure 1.2).

WP	Weather Proof	UNG	Ungrounded
RT	Rain Tight	GR	Grounded
DT	Dust Tight	R	Recessed
PS	Pull Switch	DW	Dishwasher

Figure 1.2 Symbols and Notations.

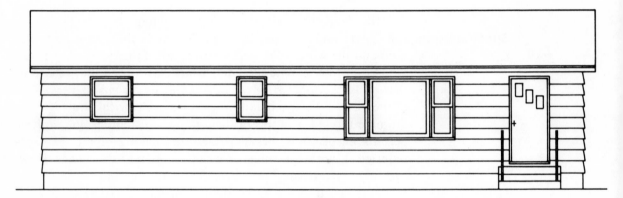

Front Elevation

(a)

Figure 1.1 The working drawing. (a) Front elevation. (b) Plot plan. (c) Detail drawing. (d) Sectional drawing. (e) Floor plan. (Courtesy of B & K Lumber Company)

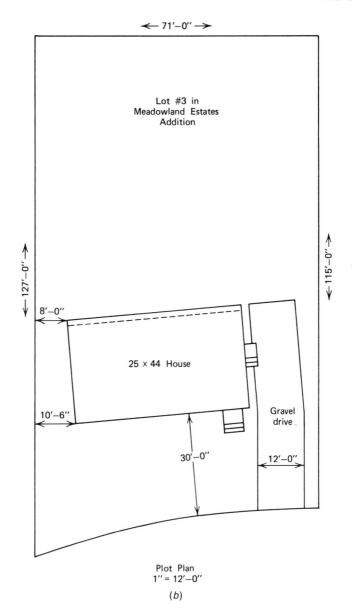

Plot Plan
1" = 12'-0"

(b)

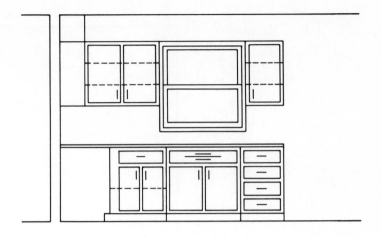

Cabinet View "A"

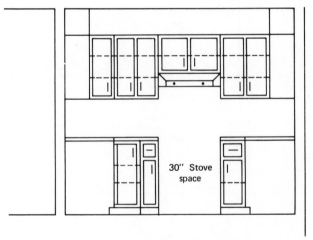

30" Stove space

Cabinet View "B"

(c)

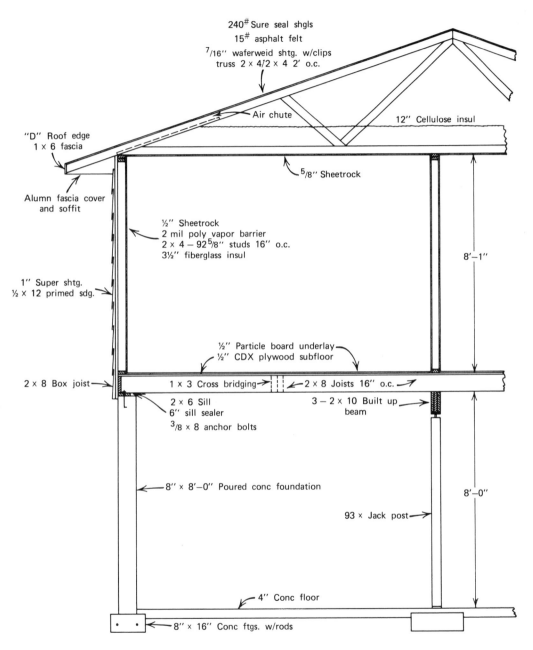

240# Sure seal shgls
15# asphalt felt
7/16" waferweid shtg. w/clips
truss 2 x 4/2 x 4 2' o.c.

Air chute

12" Cellulose insul

"D" Roof edge
1 x 6 fascia

5/8" Sheetrock

Alumn fascia cover
and soffit

½" Sheetrock
2 mil poly vapor barrier
2 x 4 – 92⅝" studs 16" o.c.
3½" fiberglass insul

8'–1"

1" Super shtg.
½ x 12 primed sdg.

½" Particle board underlay
½" CDX plywood subfloor

2 x 8 Box joist

1 x 3 Cross bridging
2 x 8 Joists 16" o.c.

2 x 6 Sill
6" sill sealer
3/8 x 8 anchor bolts

3 – 2 x 10 Built up
beam

8" x 8'–0" Poured conc foundation

93 x Jack post

8'–0"

4" Conc floor

8" x 16" Conc ftgs. w/rods

(d)

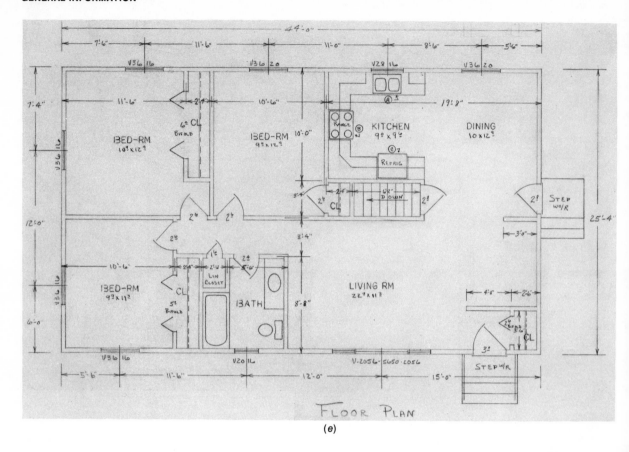

FLOOR PLAN

(e)

1-4
Electrical Standards

Because of the tremendous growth in the use of electrical energy for light, power, and heat and because there is the ever-present danger of fire or electrical shock through some failure of the electrical system, the electrician must use approved listed and labeled materials and must perform all work in accordance with recognized standards.

The National Electrical Code

The National Electrical Code is a set of rules and regulations. It is the basic standard that governs the proper method for installing electrical materials so that the completed job may be sufficiently safe to protect both life and property.

The National Electrical Code is recognized in the United States as the safety standard for the electrical construction industry.

The National Electrical Code sets forth the minimum requirements for electrical safety. However, the code book is not intended to be a design manual. It is the National Electrical Code that the electrical inspector refers to when inspecting an electrical installation. Obviously, the National Electrical Code is among the most important reference books for electricians. For convenience, the National Electrical Code will be designed as NEC throughout this book.

When the NEC is legally adopted by a town, city, or state as the official electrical code of that governmental body, it becomes law. Compliance with the code in that locality then becomes mandatory when it is officially adopted or accepted.

The purpose of the NEC is discussed in Article 90-1.

90-1 Purpose (a) The purpose of this code is the practical safeguarding of persons and property from hazards arising from the use of electricity. This NEC book is revised every three years.

Local Electrical Codes Many cities, counties, states, and municipalities adopt their own electrical code or electrical ordinances, usually based on the requirements of the National Electrical Code. Electrical installations in any locality must at least meet the requirements of the local electrical code or electrical ordinances, which are usually based on the requirements of the National Electrical Code. In some instances, these local codes include special provisions, so it is important for electricians to acquaint themselves with such local regulations as may affect them.

The responsibility of the interpretation of local codes and ordinances rests with the local inspection authority having jurisdiction in that area.

In Article 90-4 under enforcements, it is clear that the local electrical inspector has the final word in any situation.

90-4 Enforcement
The authority having jurisdiction of enforcement of the codes will have the responsibility for making interpretations of the rules, for deciding upon the approval of equipment and materials, and for granting the special permission contemplated in a number of the rules.

1-5 Permits and Inspections In many areas, regardless of population, if the locality has a building code it is usually necessary to obtain a permit from the city or county building inspection office before a wiring installation may be started (Figure 1.3).

After the rough-in of the electrical system has been completed, the electrician will call for a rough-in inspection. Walls cannot be covered until the inspector has made the inspection and signed his name to the building permit.

The utility company usually will not set the meter in the service or supply power to the house until an inspection has been made and an inspection certificate has been turned in (Figure 1.4).

In some localities, electrical installations can be legally made only by a licensed electrician. Check with the building inspection authority in your area for local rules and ordinances.

1-6 Testing Laboratories

Even though electrical wiring devices and equipment are carefully installed, the devices and equipment themselves can be hazardous unless properly designed for a specific purpose.

The National Board of Fire Underwriters' Laboratories, Inc. established testing laboratories in many areas of the United States for manufacturers to have their products, devices, or equipment tested. They submit samples to the laboratories for testing. When the product meets the minimum underwriters standards of safety, the UL label is attached (Figure 1.5).

Tested products that meet the UL safety standards are listed under various categories. The listed products that are of interest to you are contained in the Underwriters' Laboratories "Electrical Construction and Material List."

When purchasing and installing devices or equipment for your electrical installation, it is essential that you look for the UL listing label. Don't take chances with inferior materials that have not been listed by Underwriters.

Summary
1. With a working drawing, the electrician has several guidelines that tell how to proceed with the installation.
2. Specifications are a part of the building plans.
3. Symbols and notations have a standard interpretation.
4. The National Electrical Code is a set of rules and regulations that governs the methods of the installation for protection.
5. Because of the ever-present danger of fire and electrical shock, there is a need for recognized standards.
6. The electrical inspector has the final word as the inspection authority.
7. Materials approved for a specific purpose will be tested, listed, and labeled by Underwriters' Laboratories.

CITY OF ALBERT LEA DEPARTMENT OF FIRE AND INSPECTION DIVISION
BUILDING PERMIT APPLICATION

Approved By ....

Disapproved By ....

Hold ....

C-No. ....

Date ...., 19........

Owner ....

Contractor ....

Architect or Eng. ....

Elect. Contractor ....

┌─── **IMPORTANT** ───┐
BEFORE STARTING WORK CONTACT
N. W. Bell Telephone Co.____373-6441
City of Albert Lea_____373-2393
Interstate Power Co._____373-2371
Williams Bros. Pipeline__612-633-1555
For buried cables, pipes, etc.
└──────────────────────┘

Zoning Dist. .... Fire Zone ....

Address ....

Address ....

Address ....

Address ....

House Number	Street or

Lot	Block	Addition or Sub.

DESCRIPTION OF BUILDING

Front or Width Feet	Side or Length Feet	Height Feet	Number of Stories	Constructed of	Contents Cubical or Sq. Ft.	Cost of Work Covered by this Permit
						$

Structure Used As	Number of Rooms	Number of Units	Interior Finish	Kind of Heat	Basement	Garage

REMARKS

....

....

....

The foregoing is a true and correct description of the improvement contemplated by the undersigned applicant, and the applicant states that he will have full authority over the construction of same, and hereby agrees to comply with all ordinances of the City applicable to building and zoning and assumes all responsibility for such compliance. It is understood that the improvement shall not be used until Certificate of Occupancy and Compliance has been issued by the Building Inspector.

Owner or Authorized Agent

PERMIT AND RECEIPT   **X** ....

Fee_____ State Tax_____ Total_____

In consideration of the above application and the payment of $ ....permit fee a permit is hereby granted for the above described improvement conditioned upon the terms and specifications set forth above, and the faithful observance of all the provisions of the City building code, zoning ordinance and all other ordinances applicable to same. All permits issued are subject to all property restrictions.

Date Issued ...., 19...... By..
                                                                    Permit Clerk

This permit shall expire by limitation and become null and void, if the building or work authorized by this permit is not commenced within six months from the date of this permit, or if the building or work authorized by this permit is suspended or abandoned at any time after the work is commenced for a period of six months.

A Certificate Of Occupancy Is Required Before Occupancy Of All Buildings Or Change in Use Thereof
CALL FOR ALL INSPECTIONS 373-6429

TRADES

Figure 1.3 Building permit application. (City of Albert Lea, Minnesota)

GENERAL INFORMATION

Minnesota State Board of Electricity
1954 University Ave., St. Paul, Minn. 55104—Phone 645-7703

REQUEST FOR ELECTRICAL INSPECTION   P 85344

CHECK BELOW WORK COVERED BY THIS REQUEST

Type of Building	New	Add.	Rep.	Check Appliances Wired For		Check Equipment Wired For	
Home	☐	☐	☐	Range	☐	Temporary Wiring	☐
Duplex	☐	☐	☐	Water Heater	☐	Lighting Fixtures	☐
Apt. Bldg.	☐	☐	☐	Dryer	☐	Electric Heating	☐
Commercial Bldg.	☐	☐	☐	Furnace	☐	Silo Unloader	☐
Industrial Bldg.	☐	☐	☐	Air Conditioner	☐	Bulk Milk Tank	☐
Farm	☐	☐	☐	List Others Here		List Others Here	
Other_____	☐	☐	☐				

COMPUTE INSPECTION FEE BELOW

Service Entrance Size:	#	Fee	Feeders & Subfeeders:	#	Fee	Circuits:	#	Fee
0 to 100 Amps.			0 to 30 Amperes			0 to 30 Amperes		
101 to 200 Amps.			31 to 100 Amperes			31 to 100 Amperes		
Above 200___Amps.			Above 100___Amps.			Above 100___Amps.		
Transformers			Remote Control Circ.			Partial or other fee		
Signs			Special Inspection			Minimum fee $5.00		
Remarks						**TOTAL FEE**		

I, the Electrical Inspector, hereby certify that the above inspection has been made.

(Rough-in)_____ Date _____

(Final) _____ Date _____

This request void 18 months from

This request void 18 months from

P 85344

Date of this Request_____ .

I, as ☐ Licensed Electrical Contractor ☐ Owner, do hereby request inspection of the above electrical wiring installed at:

Street Address or Route No. _____City_____

Section_____ Township _____ Range_____ County _____

Which is occupied by _____
(Name of Occupant)

Is a roughin inspection required on this job?   No ☐   Yes ☐   Ready Now ☐   Will Call ☐

Power Supplier _____ Address _____

Electrical Contractor_____ Contractor's License No. _____
(Company Name)

Mailing Address _____
(Electrical Contractor or Owner Making This Installation)

Authorized Signature _____ Phone No. _____
(Electrical Contractor or Owner Making This Installation)

STATE BOARD COPY

This inspection request will not be accepted by the State Board unless proper inspection fee is enclosed.

Figure 1.4 Electrical inspection permit. (State of Minnesota)

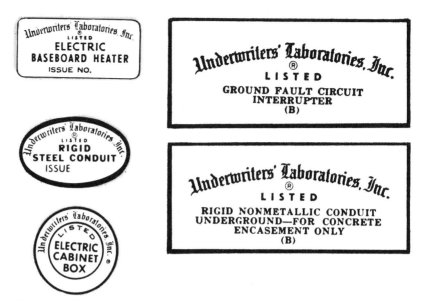

Figure 1.5 Types of labels that appear on products covered by Underwriter's Laboratories, Inc.

Problems

1-1 Explain how the architect uses a set of blueprints to convey instructions to the electrician on the job.

1-2 The electrician can find out how a house is constructed and what materials will be used from:

a. A picture of the house.

b. A set of blueprints.

c. The National Electrical Code.

1-3 Explain the purpose of the specifications.

1-4 Explain why the architect uses electrical symbols.

1-5 The National Electrical Code is a standard for the electrical industry. It is used by electricians to determine:

a. How to wire a house.

b. How to design an electrical system.

c. The minimum safety standards for installing and maintaining electrical wiring and equipment.

1-6 Explain the purpose of the NEC.

Chapter 2 Electrical Symbols and Outlets

Instructional Objectives
1. To become more familiar with electrical symbols.
2. To learn how to identify the different types of outlets, boxes, and switches used in a dwelling.
3. To understand the need for calculating the number of conductors in a box.
4. To make it possible to know how to use the NEC book to find the correct information relative to the location of outlets.
5. To become familiar with the methods of mounting the various electrical devices used in a residence.

Self-Evaluation Questions Test your prior knowledge of the information in this chapter by answering the following questions. Watch for the answers as you read the chapter. Your final evaluation of whether you understand the material is measured by your ability to answer these questions.

1. What is an outlet?
2. What symbols are used to show different types of outlets?
3. How are locations of outlets usually determined?
4. What symbol represents a ceiling outlet?
5. What symbol represents a three-way switch?
6. How can you tell, from looking at the plan, which switches control the various outlets?
7. At what height do you install convenient outlets?
8. How do you mount lighting fixtures?
9. Can you identify a recessed fixture?
10. What electrical symbol represents a wall bracket?

2-1 Electrical Symbols Electrical symbols drawn on an architectural plan are used to show the location and type of electrical devices required for the installation.

When plans are drawn, architects do their best to follow the accepted standards in representing materials and equipment. New materials are always being developed, requiring new symbols and designations. The American National Standards Institute (ANSI) and many other trade groups are working constantly to standardize procedures. Figure 2.1 shows electrical wiring symbols and their meanings.

The newcomer or apprentice to the electrical construction industry should make every effort to learn material identification in order to associate symbols with wiring devices and other electrical materials.

As stated in Chapter One, the drawing most often used by the electrician is the floor plan, because most of the wiring requirements are found on this plan. Figure 2.2 shows a simplified electrical floor plan with typical electrical outlets and switches designated by standard electrical symbols. The curved broken lines indicate which outlet the switch controls.

The National Electrical Code defines an outlet as "a point on the wiring system at which current is taken to supply utilization equipment." The term "outlet" is used very broadly by experienced electricians to include duplex receptacles, lighting outlets, switches, and similar control devices in a wiring system. However, each type of outlet is represented on the plans as a symbol (Figure 2.1).

A study of the floor plan for a single family dwelling, Figure 2.2, shows that many different electrical symbols are used to represent the various electrical devices and equipment used in residential construction.

2-2
Fixtures and Outlets

It is common practice among architects in residential construction to include in the specifications a certain amount of money as a "fixture allowance" for the purchase of the electrical light fixtures. The selection of fixtures is left up to the owner or builder. However, the electrical contractor includes this allowance in the bid. Should the fixture cost exceed the fixture allowance in the bid and specifications for an exclusive custom-built home, the owner is expected to pay the difference. For a smaller conventional house, the contractor usually purchases and hangs competitive fixtures as a part of the contract.

If the builder has made a selection of the fixtures prior to the drawing of the plans, the architect can specify these fixtures in the plans and specifications. This will provide the electrician with

USA Standard
Graphic Electrical Wiring Symbols
for Architectural and Electrical Layout Drawings

The symbols shown on this page have been taken from, or adapted from American Standards Association Standard ASA Y32.9—1962, now designated as a USA Standard.

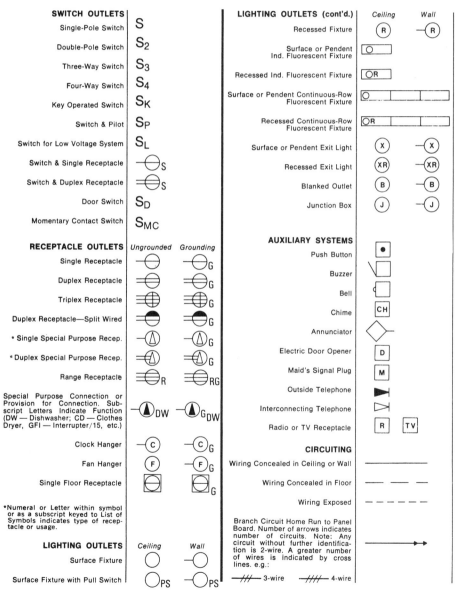

Figure 2.1 Standard electrical symbols.

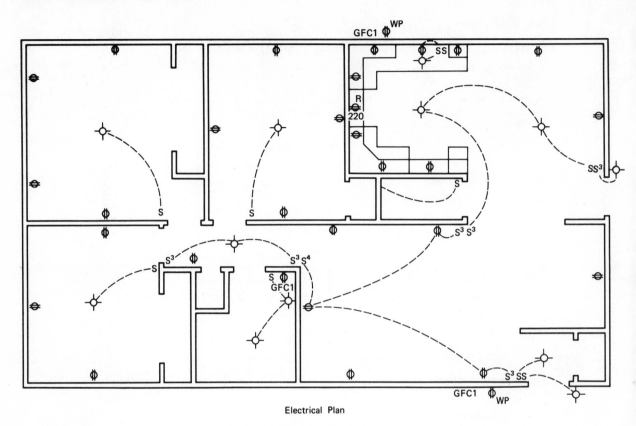

Electrical Plan

Figure 2.2 Electrical floor plan.

advance information relative to the roughing in or mounting that may be required for the fixtures. This is especially true when recessed fixtures are specified. If the fixtures are not selected in advance to construction, the electrician will usually install outlet boxes that have standard fixture-mounting studs where heavy fixtures will be anticipated. Other lightweight fixtures can be mounted either to an outlet box or box plastering using a bar strap and two 8-32 machine screws. Figure 2.3 illustrates several types of ceiling fixture outlet boxes.

**2-3
Switch Outlets** The symbol for a residential-type single pole switch is shown on the floor plan by the letter "S," a three-way switch "S_3," and a four-way switch "S_4." Figure 2.4 illustrates several switch outlet boxes. A

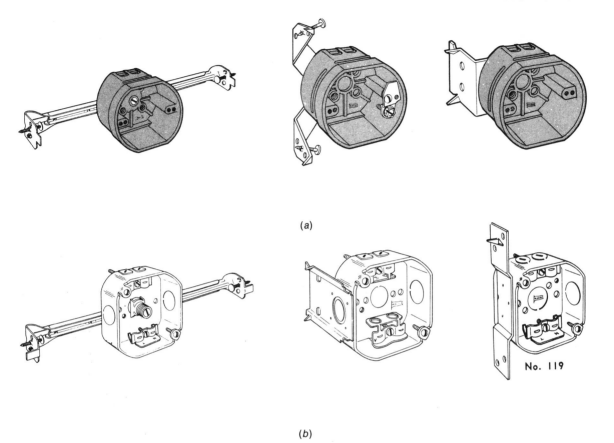

Figure 2.3 Ceiling fixture outlet boxes. (*a*) Four-inch ceiling boxes (nonmetallic). (*b*) Four-inch octagon set-up boxes (metal).

flush-type switch will fit into a standard 1¾″ x 2¾″ sectional box (either steel or plastic). If and when the plans call for two or more switches or a switch and a convenience outlet at the same location, a 2 gang box is used. Single outlet boxes can be ganged together by removing one side of each of the boxes and fastening them together.

NEC Section 370-10. In wall or ceiling. In walls or ceilings of concrete, tile, or other noncombustible material, boxes and fittings shall be so installed that the front edge of the box or fitting will not sit back of the finished surface more than ¼ inch. In walls and ceilings constructed of wood or other combustible material, outlet boxes and fittings shall be flush with the finished surface or project therefrom (Figure 2.5).

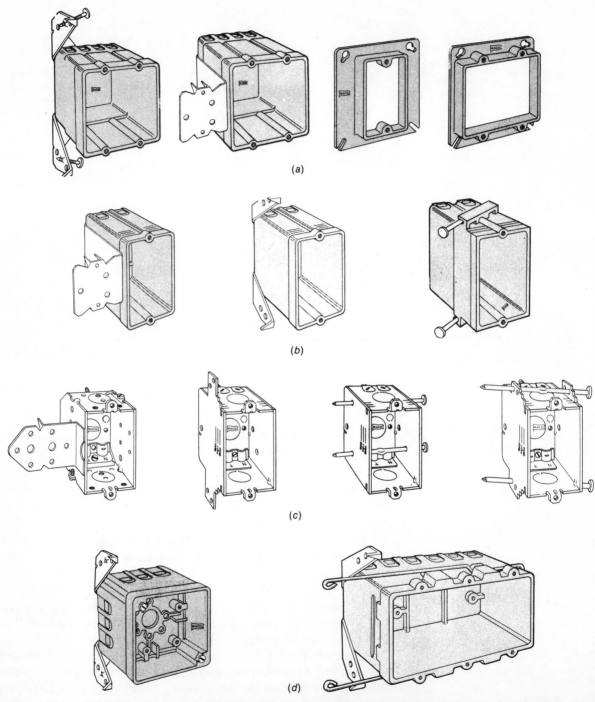

Figure 2.4 Switch outlet boxes. (*a*) Four-inch nonmetallic. (*b*) Single gang nonmetallic. (*c*) Metal. (*d*) Three gang nonmetallic.

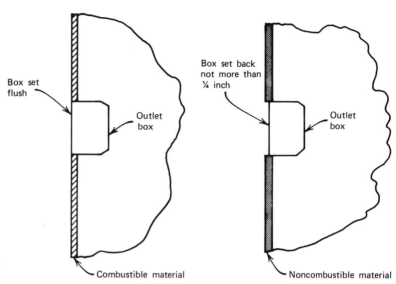

Box set flush

Outlet box

Box set back not more than ¼ inch

Outlet box

Combustible material

Noncombustible material

Figure 2.5 Switch outlet boxes installed in wall.

2-4
Junction Boxes

Junction boxes are at times placed in the circuit for convenience. A 4″ square box serves a twofold purpose: (1) as a convenient outlet box (when a single-gang plaster ring is used) and (2) as a box to make up connections. At times, junction boxes are installed in the attic for the door chime transformer or ventilation fan thermostat (Figure 2.6).

NEC Section 300-15. Boxes and Fittings—where required: (b) Box Only. A box shall be installed at each conductor splice connection point, outlet, switch point, junction point, or pull point for the connection of metal-clad cable, mineral-insulated metal sheathed cable, type AC cable, nonmetallic-sheathed cable or other cables, at the connection point between any such cable system and a raceway system and at each outlet and switch point for concealed knob-and-tube wiring.

2-5
Special Purpose Outlets

It is the responsibility of the electrician to check when a special-purpose outlet is indicated on the plans or in the specifications. It may indicate a special procedure such as a need for a polarized or grounded receptacle or a special 240-volt circuit.

Special-purpose outlets are installed in the rough-in for the dishwasher, clothes dryer, built-in cook-top, oven, and others. These outlets are usually indicated on the floor plans and are described by a notation by the symbol.

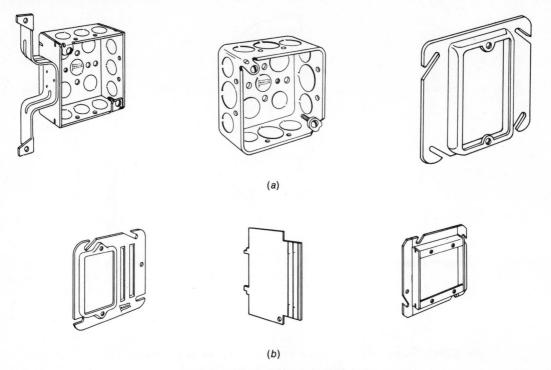

(a)

(b)

Figure 2.6 Junction boxes, plaster rings, dividers. (a) Four-inch square junction boxes. (b) Plaster rings and dividers. (Raco, Inc.)

2-6
Number of Conductors
in a Box

The number of conductors allowed in a standard-size outlet or switch box is specified in the National Electrical Code (Table 2.1—Table 370-6(a), NEC).

Section 370-6 Number of conductors in switch, outlet, receptacle, device and junction boxes. Boxes shall be of sufficient size to provide free space for all conductors in the box.

(a) Standard Boxes. The maximum number of conductors, not counting fixture wires permitted in standard boxes, shall be listed in Table 370-6(a).

(1) Table 370-6(a) shall apply where no fittings or devices, such as fixture studs, cable clamps, hickeys, switches, or receptacles are contained in the box and where no grounding conductors are part of the wiring within the box. Where one or more fixture studs, cable clamps, or hickeys are contained in the box, the number of conductors shall be one less than shown in the Tables; an additional deduction of one conductor shall be made for each strap containing one or more devices; and a further deduction of one conductor shall be made for one or more grounding

conductors entering the box. A conductor running through the box shall be counted as one conductor, and each conductor originating outside of the box and terminating inside the box is counted as one conductor. Conductors, no part of which leaves the box, shall not be counted. The volume of a wiring enclosure (box) shall be the total volume of the assembled sections, and where used, the space provided by plaster rings, domed covers extension rings, etc. that are marked with their volume in cubic inches.

The code requires that boxes other than those described in Table 370-6(a) must be "durably and legibly marked."

Section 370-6(b). Other Boxes. Boxes 100 cubic inches or less, other than those described in Table 370-6(a), nonmetallic boxes and conduit

Table 2-1
Number of Conductors in a Box (NEC Table 370-6(a): Metal Boxes

Box Dimension, Inches Trade Size or Type	Min. Cu. In. Cap.	Maximum Number of Conductors				
		No. 14	No. 12	No. 10	No. 8	No. 6
4 x 1¼ Round or Octagonal	12.5	6	5	5	4	0
4 x 1½ Round or Octagonal	15.5	7	6	6	5	0
4 x 2⅛ Round or Octagonal	21.5	10	9	8	7	0
4 x 1¼ Square	18.0	9	8	7	6	0
4 x 1½ Square	21.0	10	9	8	7	0
4 x 2⅛ Square	30.3	15	13	12	10	6
4 11/16 x 1¼ Square	25.5	12	11	10	8	0
4 11/16 x 1½ Square	29.5	14	13	11	9	0
4 11/16 x 2⅛ Square	42.0	21	18	16	14	6
3 x 2 x 1½ Device	7.5	3	3	3	2	0
3 x 2 x 2 Device	10.0	5	4	4	3	0
3 x 2 x 2¼ Device	10.5	5	4	4	3	0
3 x 2 x 2½ Device	12.5	6	5	5	4	0
3 x 3 x 2¾ Device	14.0	7	6	5	4	0
3 x 2 x 3½ Device	18.0	9	8	7	6	0
4 x 2⅛ x 1½ Device	10.3	5	4	4	3	0
4 x 2⅛ x 1⅞ Device	13.0	6	5	5	4	0
4 x 2⅛ x 2⅛ Device	14.5	7	6	5	4	0
3¾ x 2 x 2½ Masonry Box/Gang	14.0	7	6	5	4	0
3¾ x 2 x 3½ Masonry Box/Gang	21.0	10	9	8	7	0
FS—Minimum Internal Depth 1¾ Single Cover/Gang	13.5	6	6	5	4	0
FD—Minimum Internal Depth 2⅜ Single Cover/Gang	18.0	9	8	7	6	3
FS—Minimum Internal Depth 1¾ Multiple Cover/Gang	18.0	9	8	7	6	0
FD—Minimum Internal Depth 2⅜ Multiple Cover/Gang	24.0	12	10	9	8	4

bodies having provisions for more than two conduit entries shall be durably and legibly marked by the manufacturer with their cubic inch capacity, and the maximum number of conductors permitted shall be computed using the volume per conductor listed in Table 370-6(b) and the reductions provided for in Section 370-6(a)(1).

**2-7
Location of Outlets**

There are no hard and fast rules for locating most outlets, however, the electrician should verify all dimensions before starting the rough-in either by the notes on the floor plan or by the directions written in the specifications. Care should be taken in checking the kitchen and bathroom areas. Usually the kitchen and bath will have a detail drawing showing cabinets and space for equipment, sink, range, oven, and cook-top. Also check for tile, if used. Switch and duplex receptacles should be installed so that when finished they are in the tile. Check for the swing of doors and make sure the switch outlet isn't installed behind the door.

Unless otherwise noted, heights for convenience outlets and wall switches are given in the specifications and are usually from the finished floor to the center of the outlet. Proper allowances must be made when roughing-in for the kind and thickness of sheet-rock or panel used on the walls and ceiling.

Commonly specified heights for wall duplex receptacles outlets are 14 inches and for switch outlets 48 inches. Switches and receptacles over counter-tops in kitchen and bathrooms are mounted 42 inches to top of outlet.

Materials vary in thickness, so check plans and spaces.

Summary

1. The American National Standards Institute and trade groups work for standardized procedures.
2. Learn symbols and associate them with wiring devices.
3. The National Electrical Code defines an "outlet" as a point on the wiring system where current is taken to supply a circuit.
4. Why it is important to check the plans and specifications before starting the job.
5. The National Electrical Code has specific information relative to the location of outlets.
6. It is your responsibility to check special purpose outlets before starting the job.

Problems   **2-1** Draw a one-room floor plan. Locate two switch outlets and two light outlets (one ceiling and one wall bracket). Show switch to light relationship with dash lines.

2-2 Draw a sketch of a concrete wall. Show how an outlet box would be installed to the finished wall.

2-3 Draw a sketch of a combustible material wall showing how an outlet box would be installed to the finished wall.

2-4 Explain the purpose of a floor plan with reference to electrical outlets.

2-5 Draw a special outlet for a dishwasher, water heater, and clothes dryer showing proper symbols and notations.

2-6 Draw the proper symbols for the following outlets: a four-way switch, a duplex grounding type receptacle, a single pole switch, and a duplex split-circuit receptacle.

Part 2 Building Electric Circuits

Chapter 3

**Determining
the Number
of Circuits
Required**

**Instructional
Objectives**
1. To learn how to calculate the occupied floor area of a single-family dwelling.
2. To become familiar with the NEC to determine the basic requirements for branch circuits.
3. To become aware of the minimum number of lighting circuits required in a dwelling.
4. To understand the NEC relating to the number of small appliance branch circuits.
5. To learn how to determine the total load requirements in amperes for general lighting circuits.

**Self-Evaluation
Questions**
Test your prior knowledge of the information in this chapter by answering the following questions. Watch for the answers as you read the chapter. Your final evaluation of whether you understand the material is measured by your ability to answer these questions. When you have completed the chapter, return to this section and answer the questions again.

1. What dimensions are used when measuring the area of a dwelling?
2. How is the total lighting load in amperes determined?
3. What is unit load per square foot for a single-family dwelling?
4. According to the NEC, on what basis is the minimum number of receptacle outlets determined for most occupied rooms in a home?
5. What is the appliance circuit?
6. Is an unoccupied basement area included in the calculations?
7. Would you install a lighting outlet in an attic?
8. Where are receptacles to be located in the kitchen?

9. Is it necessary to install lighting outlets in every habitable room?
10. What type of circuits must be provided for receptacle outlets in the kitchen, pantry, breakfast room, and family room?

In the proceeding chapters we discussed the working drawing, specifications, symbols, and the need for safety standards. We are now ready to calculate branch circuits. It is usually standard practice for the electrician to plan for and lay out the circuits. However, they must conform to the standards established by the NEC and local code requirements.

3-1
Calculating Occupied
Floor Area
General Lighting Load

The NEC in Article 220 and section 2(b) states: Unit lighting load for dwelling units shall not be less than 3 watts per sq ft [Table 220-2(b)]. In determining this load on a watts per sq ft basis, use the outside dimensions of the dwelling (Figure 3.1). The computed floor area does not include open porches, garages, unused, or unfinished spaces unless adaptable for future use.

In section 3(d) Article 220 NEC. "Watts per sq ft" load shall be apportioned evenly among branch circuits, according to their capacitites, by using the examples in Chapter 9 for general ilumination. One 15-amp, 115-volt branch circuit would be required for every 575 sq ft of occupied floor area.

Example

3 watts per sq ft × 575 sq ft = 1725 watts ÷ 115 volts = 15 amps. A calculated load for a total occupied area of 30 ft × 55 ft = 1650 sq ft × 3 watts per sq ft = 4950 watts.

3-2
Calculating Minimum
Number of
Lighting Circuits

The total required amperes is equal to $\text{amperes} = \dfrac{\text{watts}}{\text{volts}}$

$$\text{Amperes} = \frac{4950}{115} = 43.04 \text{ amperes}$$

As a result the minimum number of circuits required for general lighting load is obtained by dividing the total required amperes by

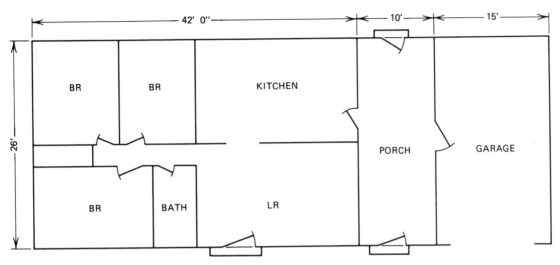

Figure 3.1 Floor plan.

the maximum amperage rating of each circuit.

$$\frac{43.04}{15} = 2.86 \text{ or } 3 \text{ circuits minimum required.}$$

Convenience Outlets In addition to general lighting load, the NEC table 220-2(b) states "receptacles other than those of the two small appliance circuits are considered part of general illumination and require no allowance for additional load."

3-3 Determining the Number of Small Appliance Branch Circuits **Kitchen** The NEC rules here, Article 220-3(b) for small appliance load, including refrigeration equipment in kitchen, pantry, breakfast room, dining room, and family room, two or more 20-ampere appliance circuits shall be provided. Such circuits shall have no other outlets. The outlets in the kitchen must be supplied by at least two 20-amp circuits. An electric clock outlet may be installed on the general lighing circuit.

NEC 220-16(a): the feeder load for the two small appliance circuits of above Article 220-3(b) is to be taken at 3000 watts or 1500 watts each two-wire circuit.

Laundry At least one additional 20-ampere branch circuit is needed to supply the laundry receptacle outlet. With no other outlets on this circuit, NEC 220-3(c), this could be calculated at 1500 watts NEC 220-16(b).

Outlets for Branch Circuits As stated in Section 3-1, convenience outlets are connected to the general lighting branch circuits with the exception of the receptacles in the kitchen (see Section 3-2). These outlets are connected to the 20-ampere small appliance circuit. Usually the location of convenience receptacles are up to the owner/builder. However, the code does give general requirements for the location of these outlets (Section 210-25(b) NEC).

Receptacles Section 210-25(b) requires that, in dwellings, receptacles must be installed:

1. in all rooms of general occupancy, no space along a wall is to be more than 6 feet from a receptacle outlet measured horizontally.
2. in any wall space 2 feet or more in width.
3. in kitchen where each counter space wider than 12 inches.
4. insofar as practicable, spaced equal distance apart.
5. with at least one wall outlet in the bathroom near basin.
6. with at least one outlet outdoors.
7. with at least one outlet in basement in addition to laundry outlet.
8. with at least one outlet in each attached garage.
9. within 6 feet of an intended location of an appliance.
10. in addition to any receptacles located or 5½ feet above the floor.

Lighting Outlets —Section 210-26(a) Dwelling Units

1. At least one wall switch controlled lighting outlet must be installed in every habitable room, bathrooms, hallways, stairways, attached garages, and outdoor entrances.
2. At least one lighting outlet to be installed in an attic, underfloor space, utility room, and basement area used for storage or space containing equipment requiring service.

3-4 Calculations Based Upon Code Sections

Area of floor space—30 ft × 55 ft = 1650 sq ft			NEC 220-2(b)
Minimum wattage, gen. lighting = 1650 × 3 = 4950 watts			220-2(b)
Two appliance circuits		3000 watts	220-16(a)
Laundry circuit		1500 watts	220-16(b)
Total Light-Appliance-Laundry		9450 watts	220-16(a) (b)

Table 220-11 demand factors
First 3000 watts at 100%   3000 watts
Remaining 6450 watts at 35%   2257 watts
Load requirements for above   5257 watts

Two—20-ampere appliance circuits   2
One—20-ampere laundry circuit   1
Three general lighting 3 15-ampere   3

                                            6 circuits

The above calculations are only minimum requirements; other circuits will be discussed in Chapter 4 and 5.

Summary 1. The minimum number of general lighting circuits will be determined by the square foot area of the floor plan.
2. The small appliance load includes the refrigeration in the kitchen and outlets in the dining and family room.
3. The appliance circuits are calculated at 3000 watts.
4. At least one 20-amp branch circuit supplies the laundry.
5. The outlets in the kitchen appliance circuit are on two different circuits.
6. Convenient outlets in bedrooms are considered part of the general lighting.
7. The NEC requires at least one outlet in every attached garage.
8. A receptacle is required in the kitchen when the counter space is 12 in. or wider.

Problems 3-1 Explain how you determine the minimum number of branch circuits for general lighting in a single family dwelling (see Figure 3.2).
3-2 Calculate the minimum number of required circuits, requirements for general lighting, and appliance circuits using the floor plan of Figure 3.3 (115/230 volts).
3-3 Draw a wiring diagram showing how to wire the appliance circuit using Figure 3.3.
3-4 Draw a wiring diagram showing the laundry circuit using Figure 3.3.
3-5 Calculate the minimum number of required circuits from the floor plan, Figure 3.3, using code requirements for general lighting, appliance circuits, and laundry circuit; indicate number of required circuits (115/230 volts).

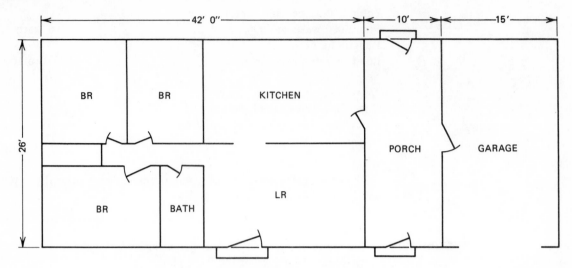

Figure 3.2 General lighting system.

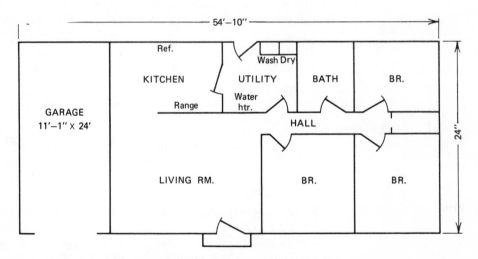

Figure 3.3 Floor plan.

Chapter 4 Branch Circuit Wiring Methods

Instructional Objectives

1. To understand why it is important to use a size conductor for the ampacity of the load required.
2. To become familiar with the type of wiring cable used in the majority of residential installations.
3. To learn why it is necessary to select the proper box for the outlet installation.
4. To learn how to install and support the cable for the outlets.
5. To understand the need for proper grounding connections when using nonmetallic cable.
6. To develop the appliance branch circuits.
7. To learn how to use the National Electric Code to determine the requirements for installing outlets and receptacles.
8. To understand the operation of a dimmer switch.
9. To learn how to install and make connections for three-and four-way switches.
10. To develop an ability to calculate the floor space of a residence for calculating branch lighting circuits.
11. To understand the need for the installation of a ground-fault circuit interrupter.

Self-Evaluation Questions Test your prior knowledge of the information in this chapter by answering the following questions. Watch for the answers as you read the chapter. Your final evaluation of whether you understand the material is measured by your ability to answer the questions. When you have completed the chapter, return to this section and answer the questions again.

1. When running type NM cable through wooden members, what precaution should be taken?
2. What is the purpose of the bare wire in nonmetallic-sheathed cable?

3. How do you support nonmetallic cable?
4. Which material has the best current-carrying capacity?
5. What is meant by ampacity?
6. What is used to determine the size of a conductor?
7. Does the insulation on a wire make any difference to its current-carrying capacity?
8. How do you determine the number of outlets on a branch circuit?
9. Does the NEC specify the number of outlets to be connected to a circuit?
10. Why does the Code limit the number of conductors in a box?
11. How are boxes grounded in a nonmetallic cable system?
12. How many travelers are needed for a three-way switch?
13. What is meant by ground-fault circuit interrupters?

4-1 Branch Circuit Wiring Methods

One of the most important parts of any residential electrical system is the conductor that makes up the wiring circuit.

Conductor Materials All residential circuit conductors consist of an insulated length of wire, usually copper. However, aluminum and copper-clad aluminum have made considerable inroads into the electrical industry in the last few years. Aluminum oxidizes if exposed to the air; this oxidation produces a layer of oxide on the surface of the conductor. If a conductor is allowed to oxidize at the terminal point or splice, the current-carrying ability of the conductor will be seriously reduced, because the oxide layer does not conduct electricity.

Copper Because copper is an excellent conductor, it is easy to make and handle and does not oxidize as much as aluminum. It is the conductor most often used in electrical installations.

Aluminum Aluminum is lighter than copper but not as good a conductor. To obtain the same current-carrying capacity, an aluminum conductor must be slightly larger than one made of copper. As stated above, aluminum oxidizes rapidly. The aluminum oxide formed acts as an insulator and reduces the flow of electrical current.

Conductor Sizes In the United States (prior to any change to metric sizing) the copper or aluminum conductor used in electrical installations is graded for size according to the American Wire Gauge Standard,

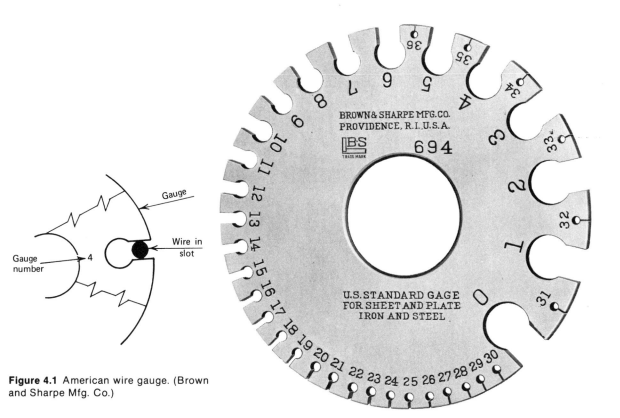

Figure 4.1 American wire gauge. (Brown and Sharpe Mfg. Co.)

simply called AWG (Figure 4.1). This gauge is used only to measure solid conductors. The outer edge of the gauge has slots that are numbered. The smallest slot into which the wire will fit is the gauge number of the wire. The wire diameter is expressed as a whole number rather than as a fractional or decimal dimension, such as Nos. 14, 12, 10, 8, and so on (Figure 4.2).

Ampacity The purpose of a conductor is to carry current from one place in a circuit to another. The size of wire used for any circuit depends on the maximum amount of current (amperes) to be carried. Ampacity is the term used to express the current-carrying capacity of a wire in amperes and depends on the material, size, and insulation of the wire.

Material The material of which the conductor is made determines how easily it will carry current. Copper is a better conductor than aluminum and will, therefore, carry more current.

Table 310-16.
Allowable Ampacities of Insulated Conductors Rated 0-2000 Volts, 60° to 90° C

Not More Than Three Conductors in Raceway or Cable or Earth (Directly Buried), Based on Ambient Temperature of 30° C (86° F)

Size	Temperature Rating of Conductor. See Table 310-13								Size
	Copper				Aluminum or Copper-Clad Aluminum				
	60° (140°F) TYPES RUW, T, TW, UF	75°C (167°F) TYPES FEPW, RH, RHW, RUH, THW, THWN, XHHW, USE, ZW	85°C (185°F) TYPES V, MI	90°C (194°F) TYPES TA, TBS, SA, AVB, SIS, †FEP, †RHH, †THHN, †XHHW*	60°C (140°F) TYPES RUW, T, TW, UF	75°C (167°F) TYPES RH, RHW, RUH, THW, THWN, XHHW, USE	85°C (185°F) TYPES V, MI	90°C (194°F) TYPES TA, TBS, SA, AVB, SIS, †RHH, †THHN, †XHHW*	
AWG									AWG
MCM									MCM
18	....	....	....	21	....	....	....	....	....
16	....	....	22	22	....	....	....	....	....
14	15	15	25	25	....	....	....	....	....
12	20	20	30	30	15	15	25	25	12
10	30	30	40	40	25	25	30	30	10
8	40	45	50	50	30	40	40	40	8

Size									Size
6	55	55	50	40	70	70	65	55	6
4	70	70	65	55	90	90	85	70	4
3	80	80	75	65	105	105	100	80	3
2	95	95	90	75	120	120	115	95	2
1	110	110	100	85	140	140	130	110	1
0	125	125	120	100	155	155	150	125	0
00	145	145	135	115	185	185	175	145	00
000	165	165	155	130	210	210	200	165	000
0000	185	185	180	155	235	235	230	195	0000
250	215	215	205	170	270	270	255	215	250
300	240	240	230	190	300	300	285	240	300
350	260	260	250	210	325	325	310	260	350
400	290	290	270	225	360	360	335	280	400
500	330	330	310	260	405	405	380	320	500
600	370	370	340	285	455	455	420	355	600
700	395	395	375	310	490	490	460	385	700
750	405	405	385	320	500	500	475	400	750
800	415	415	395	330	515	515	490	410	800
900	455	455	425	355	555	555	520	435	900
1000	480	480	445	375	585	585	545	455	1000
1250	530	530	485	405	645	645	590	495	1250
1500	580	580	520	435	700	700	625	520	1500
1750	615	615	545	455	735	735	650	545	1750
2000	650	650	560	470	775	775	665	560	2000

Correction Factors

Figure 4.2 Allowable ampacities of insulated conductors. Left column shows AWG size of conductor.

PARAFLEX® NM

60°C. NON-METALLIC SHEATH

UL FILE NUMBER E 10816

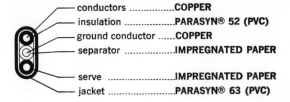

conductors ....**COPPER**
insulation ....**PARASYN® 52 (PVC)**
ground conductor ......**COPPER**
separator ....**IMPREGNATED PAPER**
serve ....**IMPREGNATED PAPER**
jacket ....**PARASYN® 63 (PVC)**

Paranite PARAFLEX® Type NM Cable with a PVC (Polyvinyl-Chloride) Jacket is designed to provide superior abrasion resistance, low temperature properties, flame retardant and crush resistant.

Type NM Cable is designed specifically for use as an internal building wire, above ground and in dry locations only. Each finished package displays the Underwriter's Laboratories Label indicating listing under the non-metallic standard.

The insulation of each conductor meets or exceeds the requirements of ASTM-D-2219, IPCEA S-61-402, and Underwriter's Laboratories Type T and TW. The jacket compound complies with the requirements of ASTM-D-1047 and IPCEA S-61-402 - 4.3.1. The copper conductors meet the requirements of ASTM-B-3 if solid and ASTM-B-8 if stranded. Paranite PARAFLEX® also meets the specific requirements of Federal government specification J-C-30A for Type NM Cable.

Size	Number of Strands	Ground Wire Size	Ampacity 30°C (86°F)	Approx. Weight Per 1000 Ft.	Approx. O.D. Inches	Package
			WITHOUT GROUND WIRE			
14/2	Solid	—	15	63	.21 x .46	250 Ft. Ctn.
12/2	Solid	—	20	78	.23 x .52	250 Ft. Ctn.
10/2	Solid	—	30	109	.25 x .55	250 Ft. Ctn.
8/2	7	—	40	185		125 Ft. Ctn.
6/2	7	—	55	286		125 Ft. Coil
14/3	Solid	—	15	89	.44	250 Ft. Ctn.
12/3	Solid	—	20	123	.48	250 Ft. Ctn.
10/3	Solid	—	30	163	.52	250 Ft. Ctn.
8/3	7	—	40	271	.70	125 Ft. Coil
6/3	7	—	55	415	.85	125 Ft. Coil
4/3	7	—	70	599	.94	125 Ft. Coil
			WITH GROUND WIRE			
14/2	Solid	14	15	72	.21 x .46	250 Ft. Ctn.
12/2	Solid	12	20	99	.23 x .52	250 Ft. Ctn.
10/2	Solid	10	30	138	.25 x .55	250 Ft. Ctn.
8/2	7	10	40	216	.39 x .73	125 Ft. Coil
6/2	7	10	55	321		125 Ft. Coil
14/3	Solid	14	15	101	.44	250 Ft. Ctn.
12/3	Solid	12	20	142	.48	250 Ft. Ctn.
10/3	Solid	10	30	196	.52	250 Ft. Ctn.
8/3	7	10	40	302	.70	125 Ft. Coil
6/3	7	10	55	443	.85	125 Ft. Coil
4/3	7	8	70	646	.96	125 Ft. Coil
2/3	7	8	95	911	1.07	500 Ft. Reel

Figure 4.3 (a) Construction of typical NEC type NM monmetallic sheathed cable.
(Wire and Cable Division, Essex International)

PAR-U-FLEX®UF
60° UNDERGROUND FEEDER CABLE

UL FILE NUMBER E 25682

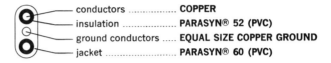

conductors .... **COPPER**
insulation .... **PARASYN® 52 (PVC)**
ground conductors ..... **EQUAL SIZE COPPER GROUND**
jacket .... **PARASYN® 60 (PVC)**

Paranite Type UF Thermoplastic Cable is recognized as underground feeder cable for direct earth burial in branch circuits by the National Electrical Code and Underwriter's Laboratories.

This cable is highly resistant to acids, alkalis, corrosive fumes, chemicals, lubricants and ground water. It is non-corroding, sunlight resistant and will not support combustion.

Paranite's PAR-U-FLEX® meets or exceeds the requirements of UL Standard 493 as well as ASTM-B-3, ASTM-B-8, and ASTM-D-2219. It also meets the specific requirements of Federal government specifications J-C-30A.

Size	Number of Strands	Insulation Thick. 64th	Insulation Thick. Mils	Approx. O.D. Inches	Ampacity (30°C) (86°F)	Approx. Weight Per 1000 Ft.	Package
			WITHOUT GROUND WIRE				
14/2	Solid	2	.030	.21 x .42	15	65	250 Ft. Ctn.
12/2	Solid	2	.030	.21 x .43	20	83	250 Ft. Ctn.
10/2	Solid	2	.030	.24 x .47	30	115	250 Ft. Ctn.
14/3	Solid	2	.030	.21 x .59	15	96	250 Ft. Ctn.
12/3	Solid	2	.030	.21 x .64	20	127	250 Ft. Ctn.
10/3	Solid	2	.030	.24 x .70	30	173	250 Ft. Ctn.
			WITH GROUND WIRE				
14/2	Solid	2	.030	.21 x .43	15	80	250 Ft. Ctn.
12/2	Solid	2	.030	.23 x .44	20	102	250 Ft. Ctn.
10/2	Solid	2	.030	.26 x .49	30	146	250 Ft. Ctn.
14/3	Solid	2	.030	.21 x .60	15	111	250 Ft. Ctn.
12/3	Solid	2	.030	.23 x .67	20	148	250 Ft. Ctn.
10/3	Solid	2	.030	.26 x .73	30	208	250 Ft. Ctn.
			SINGLE CONDUCTOR UF				
14	Solid	4	.060	.19	15	29	500 Ft. Ctn.
12	Solid	4	.060	.21	20	37	500 Ft. Ctn.
10	Solid	4	.060	.23	30	51	500 Ft. Ctn.
8	Solid	5	.080	.29	40	83	500 Ft. Ctn.
6	7	5	.080	.35	55	125	500 Ft. Coil
4	7	5	.080	.40	70	180	500 Ft. Coil
2	7	5	.080	.46	95	270	500 Ft. Coil

(b)

Figure 4.3 (b) UF underground feeder cable. (Wire and Cable Division, Essex International)

Size The larger the conductor, the more current it will carry without heating.

Insulation A conductor with insulation capable of withstanding heat will have a higher ampacity rating than a conductor of the same size with a lower insulator temperature rating. All conductors are insulated to prevent contacting each other and short-circuiting. The insulation used on building wires and cables is usually a rubber compound with an outer braid or a thermoplastic material. Table 310-13 of the NEC lists the various types of insulation available.

The NEC specifies in Section 210-19 (c) "that the minimum conductor size permitted in house wiring is No. 14 AWG."

Let us examine a branch circuit, by referring to Figure 4.12. In the drawing, the branch circuit consists of lines connecting the wall receptacle outlets in the bedrooms. In actual practice, the wiring method for this branch circuit can consist of:

Type NM monmetallic-sheathed cable, NEC Section 336. Nonmetallic-sheathed cable (NM) is used more often in residential wiring installations than any other wiring method. NM was first produced by the Rome Wire and Cable Company, which named its new product Romex. This name is still used in the trade when referring to nonmetallic cable. NM cable consists of a rubber- or plastic-insulated wire in a cloth or plastic jacket. This cable is available with two or three current-carrying conductors in sizes ranging from No. 14 through No. 2, with copper conductors, and in sizes No. 12 through No. 2 with aluminum- or copper-clad aluminum conductors. Nonmetallic-sheathed cable is available with an uninsulated conductor, called the grounding wire, which is used for grounding purposes only. It is not intended to be used as a current-carrying circuit wire (Figure 4.3).

Nonmetallic-sheathed cable is the least expensive of the various wiring methods. It is relatively lightweight and is easy to handle and install. Therefore it is widely used for residential installations.

Cable Supports NEC Section 336-5 states "type NM cable must be fastened by staples or straps so designed and installed as not to injure the cable" (Figure 4.4). The NEC requires that "a staple or strap be placed every 12 inches from every box or fitting and 4½ feet apart on the runs between the boxes."

Fastening A Cable To The Box NEC Section 370-7(c) states "where nonmetallic boxes are used, nonmetallic-sheathed cable shall extend into the box no less than ¼ inch through a nonmetallic-sheathed cable knockout opening.

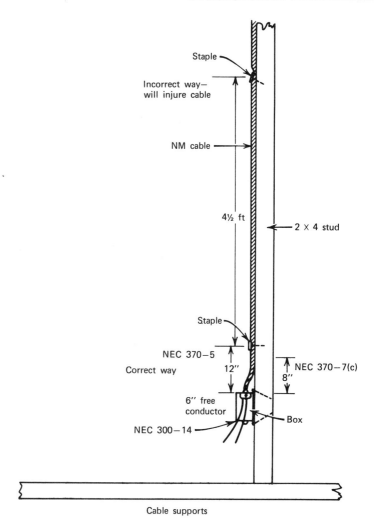

Figure 4.4 Fastening cable with staples.

When used with a single gang box and where the cable is fastened within 8 inches of the box measured along the sheath extends into the box no less than ¼ inch, securing the box shall not be required" (Figure 4.5).

Cable Protection When drilling through studs or joists, the code requires NM cable to be at least 1½" from the outer edge of the 2 x 4 or wooden member. The normal procedure is to drill a ½" or ⅝" hole in the center of the stud or joist that the cable must pass through. Sometimes it is necessary to drill the hole closer than the 1½" limit. In this case, a 1/16" steel plate is fastened to the wooden member in

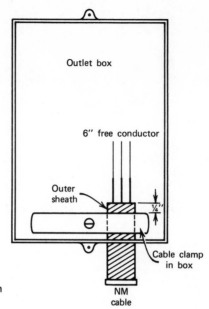

Figure 4.5 Method of fastening cable in metal box.

front of the cable to protect it from being pierced with a nail (Figure 4.6).

Not all electrical boxes are equipped with built-in clamps. Service panels and outlet boxes with ½″ knockouts require a cable connector (Figure 4.7).

NEC Section 300-14 states, "at least 6 inches of free conductor must be left at each outlet and switch point." Figure 4.8 shows a cable ripper that saves time and prevents damage to the cable during the stripping process.

4-2
Lighting Circuit
Switch Control

The flow of electrical current in the various lighting circuits of a dwelling must be controlled. This is done by a switch capable of opening and closing the circuit.

The electrician installs and connects the various types of switches used for the lighting circuits. In addition to the installation, the electrician must understand the meanings of the current and voltage ratings marked on lighting switches as well as know the NEC requirements for the installations of these switches.

Wall Switches

Switches listed by Underwriters' Laboratories, Inc., are either AC for alternating current loads or AC/DC suitable for use on either alternating or direct current circuits. There is a common miscon-

(a)

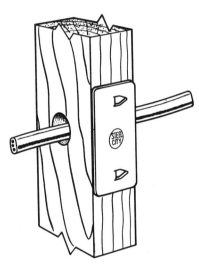

Figure 4.6 (a) Steel plate protects cable running through wood. A steel plate protector is designed to protect electrical cable or wiring raceway running through notched or bored studs, joists, rafters, or other wood structures. (b) Two integral spurs supply attachment to the stud by a hammer blow; no nails are required. (Courtesy of Midland-Ross Corporation)

(b)

ception that the AC/DC switch, being more versatile than the AC, is the superior switch. In fact, on AC loads, the AC-only switch performs much better than the dual-rated switch.

AC switches are rated for loads operating at 120 volts AC or 120/277 volts AC. Usually 120 volts AC-rated switches are intended for light-duty use and the 120/277 volt AC switches are for general or heavy-duty applications. The current rating of a switch should be adequate for the load involved.

- Now for #10-2 cable as well as #14-2 and #12-2 with/without ground, copper or aluminum
- No locknut to slow you down
- Half turn with screwdriver locks cable and connector to the box
- Locating tab shows you connection is secure
- Connection completed outside the box
- Listed by Underwriters' Laboratories, Inc.
- Patent Number 3788582

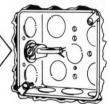

In old work, connection is made inside outlet, junction or panel box.

No. 4711. Made of high impact, self extinguishing plastic.

INSTALLATION IS EASY

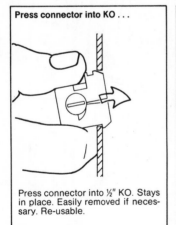

Press connector into KO . . .

Press connector into ½" KO. Stays in place. Easily removed if necessary. Re-usable.

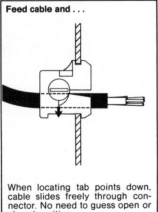

Feed cable and . . .

When locating tab points down, cable slides freely through connector. No need to guess open or closed position.

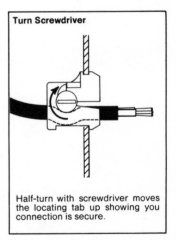

Turn Screwdriver

Half-turn with screwdriver moves the locating tab up showing you connection is secure.

Figure 4.7 Cable connector. Remove the knockout snap in the connector, insert cable. Half turn with screwdriver secures cable from pull-out. (Raco, Inc.)

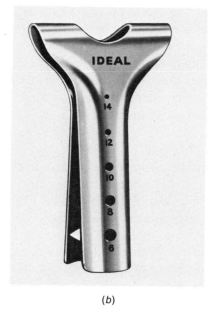

(b)

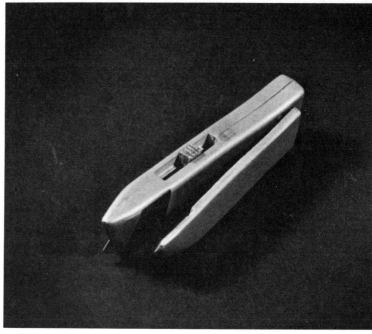

(a)

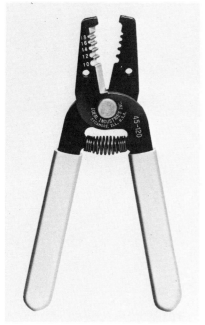

(c)

Figure 4.8 Cable strippers. (a) Grip stripper utility cable knife. (Courtesy of Blackhawk Industries) (b) Cable ripper with wire gauge. Cuts nonmetallic sheathed duplex of lead covered cable. Squeeze onto cable and pull at least six inches for cut. (Courtesy of Ideal Industries, Inc.) (c) Handy flat wire strippers; strips sizes No. 18 to No. 10 wire easily. (Courtesy of Ideal Industries, Inc.)

Toggle Switch The most frequently used switch in residential lighting circuits is the toggle switch. Toggle switches are available in three types.

Single Pole A single-pole toggle switch is identified by its two terminals and the words ON and OFF on the handle. It is usually installed when a light or group of lights must be controlled from one switching point. This type of switch is connected in series with the ungrounded (hot) wire feeding the load (Figure 4.9).

Three-Way Switch A three-way switch can be distinguished from a single-pole switch by noting that a three-way switch has three terminals. One of these terminals, the common terminal, is a darker color than the other two traveler wire terminals, which are natural brass color.

Three-way switches are used when a load must be controlled from two different switching points. Two three-way switches are used, one at each switching control point (Figure 4.10).

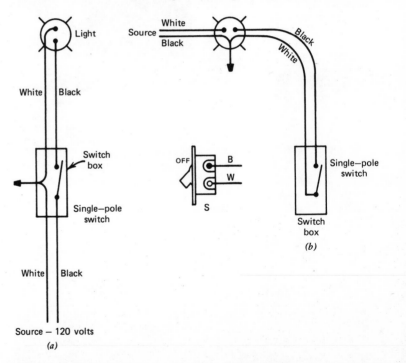

Figure 4.9 (a) Typical application of a single-pole switch to control a light from one switching phase. One hundred twenty volts feeds through switch, black wire is broke at the switch, white wire is fastened together with a wirenut. (b) With feed at light. The 120 volts feeds directly to the light outlet; a two wire cable with black and white wires is used as a switch loop between the switch and the load. The white wire is connected to the black wire at the source. NEC. 200-7.

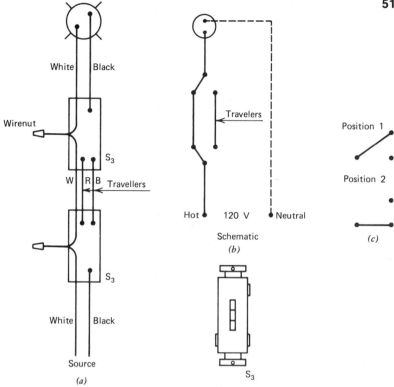

Figure 4.10 (a) Two three-way switch control, feed at the first switch control point. (b) Schematic wiring diagram. (c) The internal design of the three-way switch allows current to flow through the switch in either of its two positions.

Four-Way Switches A four-way switch can be identified readily by its four terminals and the fact that the toggle does not have ON or OFF on the toggle. The four-way switch has two positions; neither of these positions is ON or OFF. The internal mechanism of the four-way switch is constructed so that the switching contacts can alternate their positions as shown in Figure 4.11. Care must be used in connecting the traveler wires to the proper terminals. Notice the color of the terminal screws when making connections.

Some suggestions for using wall switches include:

a. Point-of-use control helps conserve energy and adds flexibility for the owner. All loads should be designed to be turned off when not in use.

b. Pilot-lighted switches offer ON/OFF indication at the switching point whenever the switch is out of sight of the load.

c. Lighted switches aid persons entering dark areas, particularly if they are unfamiliar with the room, and make sense in bathrooms.

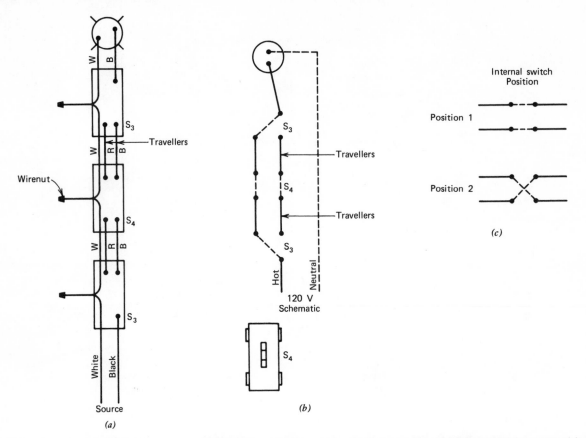

Figure 4.11 (a) Cable wiring diagram showing three-way and four-way switch control. (b) Schematic. (c) Internal switch position.

4-3
Lighting Branch Circuits
Number of Outlets Per Circuit

The planning of circuits in most residential installations is usually left up to the judgment of the electrician. There are many possible combinations or groupings of outlets. However, the code places no limits on the number of outlets that may be connected on one branch lighting circuit. There are few guidelines other than the National Electrical Code or your local electrical code on which to base the choice of particular outlets for a particular circuit.

Estimating Wattage For Outlets

When planning circuits, the electrician must consider the types of fixtures that may be used at the various outlets. Remember that symbols are used on circuit diagrams to indicate the type of each device and outlet and where they are located on the plans.

The Code does not specify the number of outlets to be connected to a circuit. However, NEC section 220-2(a) does state that "the continuous load supplied by a branch circuit shall not exceed 80 percent of the branch circuit rating." If you figure a 15-ampere, 115-volt branch lighting circuit, this means that the continuous load must be no more than 1380 watts.

$$15 \text{ amperes} \times 115 \text{ volts} \times 0.80 = 1380 \text{ watts}$$

The 15 ampere circuit is wired with No. 14 wire and protected by a 15-ampere fuse or circuit breaker. The receptacles cannot be rated at any more than 15 ampere, which means that only the ordinary household variety of receptacle may be used. However, no appliance used on the circuit may exceed 12 amperes (1380 watts) in rating.

The electrician should keep in mind that, in rooms in which the outlets are located, it does provide indications as to their possible uses, and that all circuits should be planned accordingly.

**4-4
Lighting Branch Circuit For
Bedrooms (Figure 4.12)** Every bedroom should be provided with general illumination controlled by a wall toggle or dimmer switch. Some owner-builders prefer ceiling fixtures; others prefer split receptacles, half being switched or controlled by the dimmer switch for controlled lighting, while the other half is a hot receptacle.

Review Section 2-7, Location of Outlets, page 26. Switches are usually installed near the door. The outlet box should be nailed to the stud so that the top of the switch box is 48 inches from the sub-floor. Receptacle outlet boxes should be nailed to located studs 14 inches to top of box from sub-floor.

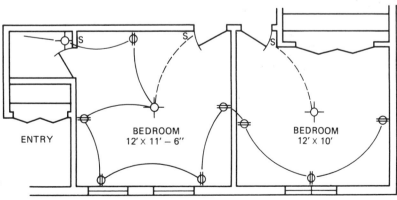

Figure 4.12 Typical circuitry of bedroom lighting and receptacles.

Remember, if using nonmetallic-sheathed cable that contains an extra conductor, this conductor is used only for grounding the box. NEC Section 210-7, "all receptacles installed on 15-ampere and 20-ampere branch circuits must be grounding type receptacles."

4-5 Bathrooms The electrician must follow very carefully the NEC requirements for the installation of lighting and receptacles installed in bathrooms.

NEC Section 210-25(b) requires at least one receptacle outlet near the basin (Figure 4.13). The Code further refers to Section 210-8(a), "all 120 volt, 15- and 20-ampere receptacles installed in the bathroom shall have ground-fault circuit interrupter protection for personnel."

What is a GFCI? A Ground Fault Circuit Interrupter, Figure 4.14, is an electronic device designed to protect people against ground fault (current leakage) before it can do any damage. Ground Fault occurs when a person comes in contact with a hot or live line (wire). This is possible either by touching an exposed wire or by simply operating a faulty appliance or power tool having a wire defect that causes the metal housing of the product to become electrically alive.

Why a GFCI? As a result of marked increases in the use of appliances (shavers, hair dryers, whirlpool pumps, etc.) and tools used by the average American homeowner, the NEC made ground fault protection mandatory in many places and situations where it was not pre-

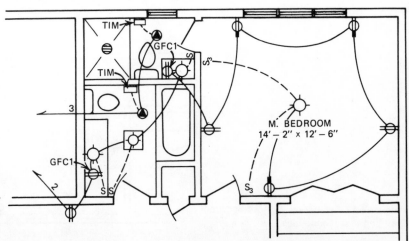

Figure 4.13 Typical circuitry of master bedroom, full bath, half bath, and hall lighting and receptacles, including ground-fault in bathrooms.

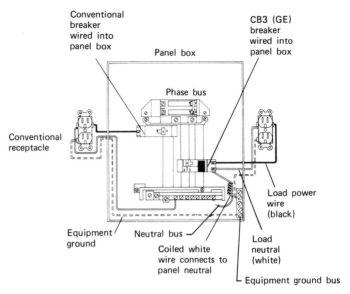

Figure 4.14 Ground-fault receptacle.

All connections to the CB3 circuit breaker are similar
to those of conventional breakers except for the
addition of a neutral connection.

viously required. Ground Fault (leakage current) is dangerous
enough to cause electrocutions, electrical fires, and other serious
accidents in the home.

How Does a GFCI Work? The GFCI continuously monitors the current in the two conductors of a circuit, the hot wire and the neutral wire. These two
currents should always be equal. If the GFCI senses a difference
between them of more than 5±1 milliamperes, it assumes the difference is ground fault current and automatically trips the circuit.
Power is interrupted which 1/40 of a second or less, which is fast
enough to prevent injury to anyone in normal health.

NEC Section 215-9 indicates that a ground fault circuit breaker
could be used to protect the feeder supplying 15- and 20-ampere
receptacle branch circuits in lieu of the GFCI receptacle. However,
the GFCI receptacle is usually located only a few feet from the
appliance being used. People are often afraid to test a circuit
breaker device located in a basement or reset it whenever it trips.
Underwriters' Laboratories require that all GFCI's be tested
monthly, so perhaps you'd rather install a GFCI receptacle in the
bathroom.

Remember, fuses and circuit breakers are designed to protect
wiring and appliances. A GFCI receptacle senses only ground fault

and is designed to interrupt at a much lower current flow to protect lives (Figure 4.15).

Bathroom Lighting To provide enough light for shaving and putting on makeup, a lighting fixture is usually installed over the mirror (controlled by a wall switch) in small bathrooms. This is enough light. However, in larger bathrooms a ceiling fixture is needed beside the light located near the mirror or medicine cabinet. For luxury and convenience, install an infared heat lamp and exhaust fan.

4-6
Outlets in Closets The Code does not require a light in the clothes closet. However, should one be installed, NEC Section 410-8 has specific rules about the placement and type of lights installed.

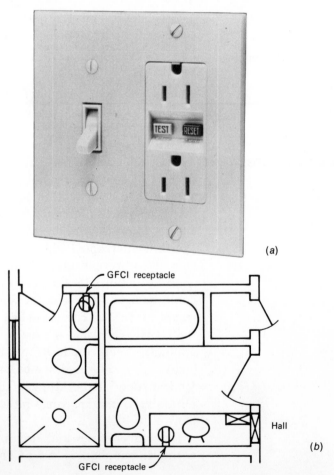

(a)

(b)

Figure 4.15 (a) Ground-fault receptacle located in each bathroom. Notice that the receptacle is duplex, and that it provides indication, reset, and test features. Switch could be used to control light over mirror, (b) GFCI wiring diagram. (Courtesy of Pass and Seymour, Inc.)

Fixtures in closets shall be installed on the wall above the door or on the ceiling over the area unobstructed to the floor, so that clearance of stored combustibles is at least 18 inches. A flush recessed solid-lens fixture is considered outside the closet area. A pendant light (a light socket on the end of a drop cord) is not permitted. 410-8(b) states, "Pendants shall not be installed."

4-7
Living Room Circuit
(Figure 4.16)
Be generous with wall receptacle outlets in the living room. Install enough receptacles so that lamps can be placed anywhere without using an extension cord.

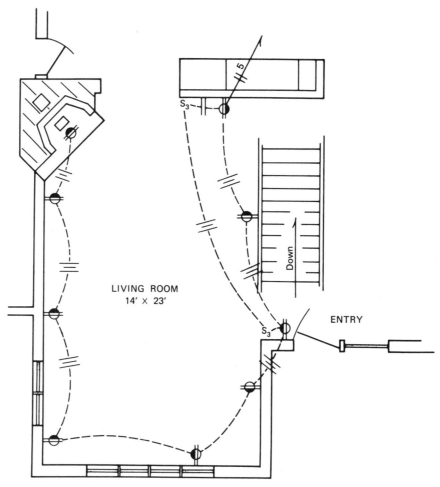

Figure 4.16 Typical living room circuitry, showing split receptacle, half hot, other half switched.

The living room is an excellent area for controlled lighting. The duplex receptacle is really two receptacles in a single device, so that two different things can be plugged in at the same time. Plan and wire the receptacles so they are split; that is, half of the receptacle is hot all of the time; the other half is switched with a dimmer switch (Figure 4.17) located at a convenient location on the wall. Three- or 4-way switches and dimmer should be installed for convenience.

In a three-wire installation, extreme care must be exercised when connecting the three wires in the device. Red wire will be the switch leg to each switched receptacle, black wire to hot receptacle, and white wire to common or neutral (Figure 4.18).

Figure 4.17 Rotary dimmer switch mounts in any standard electrical switch box and can replace any standard flush mounted switch; standard incandescent light bulbs are turned on and off by pressing the control knob, level of light is obtained by rotating the control knob. (Ideal Industries, Inc.)

Figure 4.18 Method of connecting convenience receptacles in living room. Duplex receptacles are designed so that the bridge linking the two terminal points, one on each side, can be removed. This electrically separates the two portions of the receptacle.

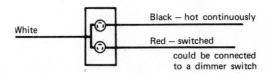

White

Black — hot continuously

Red — switched

could be connected
to a dimmer switch

4-8 The hall, entry hall, and outside lights are grouped into a circuit.
Hall Front Entry Ceiling lights should be installed in all halls. Should the halls be
and Outside rather long, two or more ceiling lights or recesed fixtures should be
installed and controlled by 3-way switches at each end. Figure 4.19
shows that at the front door two 75-watt high-boy recessed fixtures
are installed with a 150-watt recessed fixture near the center of the
hall, controlled by 3-way switches. A receptacle is installed for
convenience in the hall for a vacuum cleaner outlet. The outside
light could be a decorative bracket fixture or a ceiling fixture
installed in the overhang. Outside is a ground-fault receptacle with
a water-tight cover. See Section 4-5, Bathrooms, for ground-fault
circuit interrupter protection.

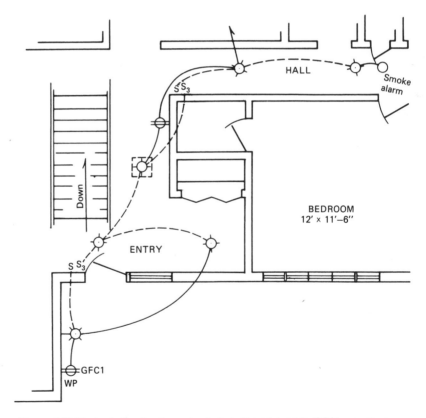

Figure 4.19 Typical circuitry for entry hall, hall, and outside GFCI receptacle.

4-9
Kitchen and Dining Area
Lighting (Figure 4.20)

For general lighting in the kitchen, there should be ample ceiling fixture outlets controlled by wall toggle switches located at each entrance to the kitchen. The prime requirement for good lighting

Figure 4.10 Typical lighting/switch circuitry in kitchen and dining room.

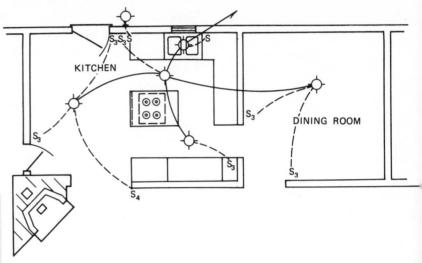

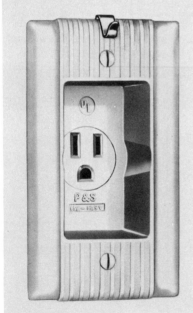

Figure 4.21 This outlet supports a clock. The receptacle is in a "well" so that the cord and plug of the clock are completely concealed. (Pass and Seymour, Inc.)

design is to eliminate shadows. A light over the sink is essential, for without this light, a person would be standing in a shadow at this location. Control this light with a wall switch located near the sink. Perhaps you could gang two switches together here, one for the light, the other for the garbage disposal under the sink. (Garbage disposal will be discussed in Chapter 5—Section 7.)

A clock-hanger type receptacle, Figure 4.21, could be installed on either a lighting circuit or on the 20-ampere small appliance circuit.

Lcoate and install a ceiling outlet in the eating area. Care should be taking in locating the outlet over the breakfast table, which is also controlled by 3-way and 4-way wall switches.

When locating the ceiling fixture outlet in the dining area, try to visualize a table located in the room. Center the outlet where you think the table would be and not at the center of the room. Control this light by 3-way wall switches or control lighting by installing dimmer switches.

4-10
Kitchen Small Appliance
Circuits (Figure 4.22)

The kitchen needs lots of receptacles. The NEC recognizes this need and requires two 20-ampere small appliance circuits to be installed in this area. See Chapter 3-3, Determining the Small Appliance Circuits.

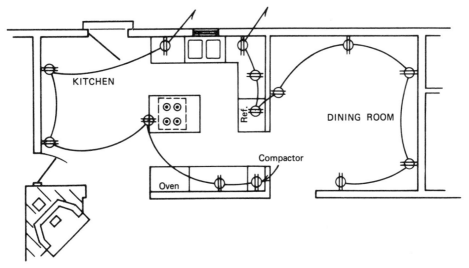

Figure 4.22 Typical small appliance circuitry, with two circuits in the kitchen.

4-11
Garage (Figure 4.23) There are some changes in the code for receptacles in the garage. NEC 210-8 "All 120 volt 15-20 amp receptacles installed in the garage shall have ground-fault circuit interrupter protection." This includes the outlet for an automatic garage door opener even though it is located on the ceiling.

For convenience, garage lights are controlled by 3-way and 4-way wall switches located at the entrance doors.

The outside bracket lights are controlled by 3-way switches, one switch located inside the breakfast room and the other near the garage door.

Installing three fixture outlets makes for good lighting and eliminates shadows between the automobiles.

The outside receptacle is on GFCI and has a weatherproof cover.

4-12
Basement The requirements for the basement vary. It depends on how elaborate the finish in the basement is to be. However, we are to have one side of the basement area as a finished family room. This will be wired like the living room upstairs and will provide for plenty of receptacles. Because the ceilings are lower in the basement, it is recommended that recessed fixtures be installed, controlled by wall switches or dimmer switches for controlled lighting.

The rest of the basement will be wired as ceiling lights are required; each storage area requires a light.

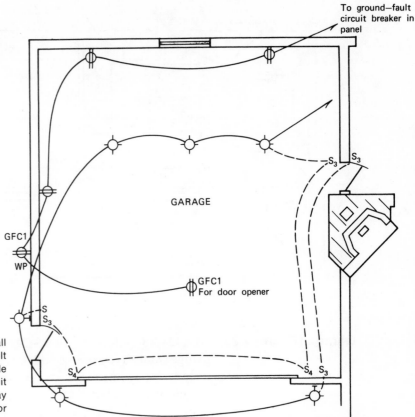

To ground—fault circuit breaker in panel

GARAGE

GFC1

WP

GFC1
For door opener

S₃ S₃

S
S₃

S₄ S₃

S₄

Figure 4.23 Garage circuitry showing all garage receptacles on a ground-fault breaker in panel. There is an outside GFCI with waterproof cover. Light circuit is controlled by three-way and four-way switching. Notice that the automatic door opener is a GFCI.

The appliances — washer, dryer, hot water heater, water conditioner, etc — will be installed here. See Chapter 5, Special Purpose Outlets.

The washer requires a circuit.
The clothes dryer requires its own circuit.
The water heater is on a separate circuit.
If a furnace is installed, some inspectors require a separate circuit.

A light to illuminate the stairs should be controlled by a 3-way switch at the top and bottom of the stairs.

Summary   1. Copper is usually the material used in residential circuit conductors.
2. Conductors used in electrical installations are graded for size according to the American Wire Gauge Standard.
3. The purpose of a conductor is to carry current.

4. Ampacity is the term used to express the current-carrying capacity of a wire in amperes.
5. Nonmetallic-sheathed cable is used more often in residential wiring than any other wiring method.
6. The NEC requires nonmetallic-sheathed cable to be supported.
7. The flow of electrical current in lighting circuits must be controlled by a switch.
8. The most frequently used switch is the toggle switch.
9. The planning of electrical circuits is usually left up to the judgment of the electrician.
10. Care must be taken when installing closet lights.
11. A ground-fault circuit interrupter is a device to protect people against shock.
12. Dimmer switches are often installed to control lighting.
13. Be generous with receptacles in the living room, kitchen, and dining area.
14. Room lights are usually controlled by wall switches near the entrance of the room.

Problems   **4-1** Draw a one-line schematic of a ceiling fixture controlled by two 3-way wall switches; hot wire is fed into the fixture outlet.

4-2 Plan and draw the layout of outlets for the small appliance circuits for the floor plan shown in Figure 4.24.

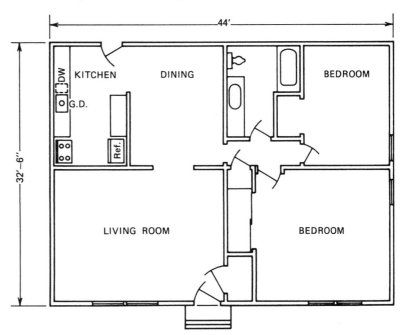

Figure 4.24 Floor Plan for Problem 4-2.

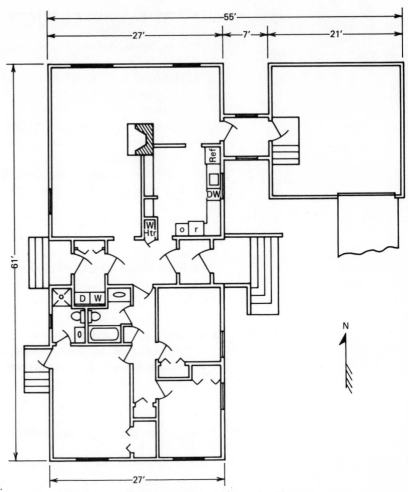

Figure 4.25 Floor plan for Problem 4-3.

4-3 Plan and draw the wiring diagram to a convenient scale for the dwelling shown in Figure 4.25.

4-4 Explain how a ground-fault circuit interrupter works?

4-5 Draw a sketch of a lighting fixture being controlled by two 3-way switches and one 4-way switch; the hot wire is fed to one of the 3-way switches.

4-6 Draw a floor plan of a house and show the dimensions; from the dimensions calculate the minimum number of light circuits needed.

Chapter 5

Special-Purpose Outlets

Instructional Objectives

1. To learn how to interpret symbols on the electrical floor plan for ranges, counter cooking units, ovens, and other major appliances.
2. To develop the ability to determine any special installation requirements for special-purpose outlets.
3. To understand the grounding of all appliances according to the Code regardless of the wiring method used.
4. To learn how to select the proper conductor size for wiring special-purpose outlets.
5. To understand the need for selecting the proper overcurrent device based on the amperage rating of the device.
6. To develop the ability to calculate the dryer circuit.
7. To understand the National Electrical Code requirements for all special-purpose outlets.
8. To learn how to use service entrance cable to connect large appliances.

Self-Evaluation Questions

Test your prior knowledge of the information in this chapter by answering the following questions. Watch for the answer as you read the chapter. Your final evaluation of whether you understand the material is measured by your ability to answer these questions. When you have completed the chapter, return to this section and answer the questions again.

1. What determines the voltage of an appliance?
2. Would you calculate the load of a range the same as you would a counter-mounted cooking top and wall oven?
3. Are most hot water heaters considered a continuous duty load?
4. How many outlets are permitted on the laundry circuit?
5. Are clothes dryers basically a 240-volt appliance?
6. How are clothes dryers grounded?

7. Is it possible to use a split receptacle for the garbage disposer and dishwasher?
8. Do the receptacles in the garage have to be a ground-fault circuit interrupter?
9. Is it permissible to install an electric wall heater in the bathroom?
10. Will the NEC permit a cord on the garbage disposal?

**5-1
Types of Branch Circuits**

The NEC recognizes two types of branch circuits. The first is the ordinary circuit serving two or more outlets consisting of permanently connected appliances, light fixtures, or receptacles for portable loads such as mixers, lamps, vacuum cleaners, and other small hand appliances, as already discussed in Chapter 3. The second type is a circuit serving a single current-consuming appliance or similar load, such as a countermounted cooking unit, oven, range, water heater, dishwasher, garbage disposer, or trash compactor using a single circuit.

Some local codes require the furnace to be on a separate circuit regardless of the horsepower rating of the blower motor.

Circuits for appliances may be either 120 or 240 volts, depending on the name-plate of the appliance.

The NEC is not clear as to when an appliance requires a branch circuit. In general, you will not only meet the requirements but will also provide the homeowner with a better installation if you provide a separate circuit for the following appliances.

**5-2
Counter-mounted
Cooking Unit**

The electric range in many areas has been replaced by the counter-mounted cooking top and built-in wall oven or by the microwave oven and radar range. The cooking top and oven are like separate units of a complete range divided for the sake of kitchen convenience, ease of operation, or kitchen appearance.

The two units are connected to individual branch circuits of lower rating than a single range unit. Normally, the range circuit terminates in a heavy-duty flush- or surface-mounted 50-ampere receptacle. The range should be equipped with a three-wire cord and plug (Figure 5.1).

A range is actually a multiunit load assembled as a single appliance. The watts rating of the range is the total of the oven load plus the total load of all of the heating elements turned to their highest

Figure 5.1 50-ampere range receptacle, surface mounted. (Courtesy of Pass and Seymour, Inc.)

setting. However, it isn't likely that all of the elements will be used at the same time. Therefore, the NEC Table 220-19 allows a demand factor for electric ranges, wall-mounted ovens, counter-mounted cooking units, and other cooking appliances over 1¾ kW rating.

A load of only 8 kW (8000 watts) may be used for any range rated at not more than 12 kW (12,000 watts). If the range is rated at over 12 kW, start with 8 kW. For each kilowatt or fraction thereof above the 12 kW, add 400 watts (or 5% of 8000 watts).

Example:

A range might have a rating of 16 kW, which exceeds 12 kW by 4 kW.

Add 400 watts for each of the extra kilowatts, or 1600 watts together.

$$8000 \text{ watts} + 1600 \text{ watts} = 9600 \text{ watts}$$
$$9600 \text{ watts} \div 240 \text{ volts} = 40 \text{ amperes}$$

The range should be wired for 40 amperes rating.

In other cases, the name-plate rating of each item, oven, and cook top must be used separately, and no demand factor may be used. It is much simpler and better to install a separate branch circuit for each cooking top and each wall oven. Rough in according to the size of load of each unit, leaving enough wire stubbed out of the wall to make connections direct to the top and oven when installed at the finish.

Microwave ovens usually draw a load of 1500 watts. A separate circuit should be roughed in when the owner intends to use this appliance.

**5-3
Room Air Conditioners**

An air conditioner is considered a room air conditioner if it is installed in the room it cools (in a window or in an opening through a wall), if it is single-phase, and if it operates at not over 250 volts.

Air conditioners are evaluated in two ways:

(a) "Cooling capacity" is the amount of heat (measured in BTU's) that an air conditioning unit can remove from the air in an hour. One BTU is the amount of heat needed to raise the temperature of one pound of water one degree Fahrenheit. This is about the amount of heat generated by a wooden match burned completely to ashes. A 12,000 BTU unit will therefore remove 12,000 BTU's of heat from an area every hour.

The capacity required for a home depends on many things, including area of the country, size of the space to be cooled, and the number of occupants in the home. However, as a general rule, 18 BTU's removed per hour will cool about one square foot in a normal home.

(b) "Energy Efficiency Ratio" (EER) is the number of BTU's of heat that one watt of electrical energy will remove from the air in one hour. The EER is determined by dividing the capacity in BTU's per hour of a unit by its power required in watts. This information is usually found on the name-plate of the unit. The EER will be a number ranging from 4.7 to 12.2.

Example:
A 10,000 BTU unit requiring 2000 watts has an EER of 5. Purchase units with high EER's; 6 to 7 is fair, but you can do better; 8 to 9 is good; and 10 or over is great.

Installation Requirements These are outlined in NEC Section 440. The air conditioner may be connected by a cord and plug. A unit switch and overload protection are built into the unit. The disconnecting means may be the plug on the cord or the manual control on the unit if it is readily accessible and not more than 6 feet from the floor, or a manually controlled switch installed where readily accessible and in sight of the unit. If the unit is installed on a circuit supplying no other load, the ampere rating on the name-plate of the unit must not exceed 80 percent of the circuit rating. Should it be installed on a circuit also supplying lighting or other loads, it may not exceed 50 percent of the circuit rating. Cords must not be longer than 10 feet if the unit operates at 120 volts and not over 6 feet if it operates at 240 volts.

**5-4
Hot Water Heater** The hot water heater is the second largest consumer of energy in the home. Therefore, much thought must be given to this appliance.

The consumption of hot water is linked to the number of people in the family, their ages, and individual habits. Oversizing wastes energy, since you are maintaining a supply of unnecessary hot water. For the average four-member household, an 80-gallon low-power heater is recommended.

Locate the heater close to the point where it is needed. Some contractors suggest several smaller heaters in larger homes, with each placed near a location needing hot water.

The power consumed by a water heater varies greatly, from 700 watts to 5000 watts. A No. 12 wire with an ampacity of 20 amperes will carry 4800 watts and is suitable for most heaters. However, some inspectors might require a No. 10 AWG.

If your service equipment consists of circuit breakers, provide a 20-amp 2-pole breaker for the heater; provide a 30-amp if you use No. 10 wire.

The following code sections apply to the load demands on a branch circuit by a water heater.

NEC 422-14(b) states that "all fixed water heaters having a capacity of 150 gallons or less shall be considered a continuous duty load."

NEC 422-5(a) Exception 2 states that for a continuous loaded appliance, "the branch circuit rating shall not be less than 125 percent of the marked rating" of the appliance.

**5-5
Clothes Dryers** Dryers are basically 240-volt, single-phase appliances, although most have a 120-volt motor-operated drum that tumbles the clothes as the heat evaporates the moisture.

A separate circuit must be provided for the electric clothes dryer because this appliance demands a comparatively large amount of power.

NEC Section 250-45(c) states "clothes dryers must be grounded," but may be grounded to the neutral of the circuit wire if the grounded conductor is No. 10 AWG or larger.

NEC 220-18 requires a circuit with a minimum of 5000 watts capacity or a capacity based on the rating of the dryer, whichever is higher. Dryers are usually installed by using a cord and plug and a receptacle, which must be 30-amperes 3-wire. The 120/240 volt type is shown in Figure 5.2.

Use any wiring method you choose. The NEC permits you to use service-entrance cable with a bare neutral as in the case of the electric range. However, it must start from the branch circuit overcurrent protection in the service equipment. The plug and receptacle will serve as the disconnecting means.

The nameplate on the dryer specifies the minimum circuit size and the maximum rating of the circuit overcurrent device.

The washer requires a separate laundry circuit [NEC 220-16(b)].

Figure 5.2 30-ampere dryer receptacle, surface mounted. (Courtesy of Pass and Seymour, Inc.)

5-6
Dishwasher A built-in dishwasher is a motor-driven appliance and should be connected to a separate 20-ampere circuit. Usually the dishwasher is located under the drainboard near the sink. Should this be the location, the NEC now states that a built-in dishwasher and trash compactor intended for use in a dwelling, if it is provided with a three-conductor cord terminated with a grounding-type attachment plug, shall be permitted if the cord is at least 3 feet to 4 feet in length and if the receptacle is located to avoid physical damage to the cord and located in an accessible area in the space occupied by the appliance or adjacent thereto. A split receptacle could be used for both the garbage disposer and the dishwasher when located under or near the sink (Figure 5.3). If the dishwasher is installed more than 4 feet from the disposer, another duplex receptacle could be roughed-in as the above directions state.

5-7
Garbage Disposer Many different types of garbage disposal units are available. However, most units are furnished with a junction box or wiring space for wiring conductors to terminate.

Since the garbage disposer is a motor-operated appliance, the NEC requires running overcurrent protection not to exceed 125 percent of the full-load current rating of the motor.

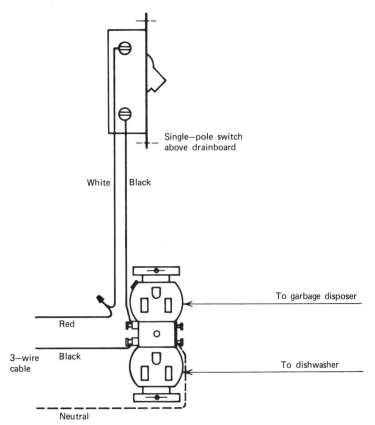

Single—pole switch
above drainboard

White | Black

To garbage disposer

Red

3—wire Black
cable

To dishwasher

Neutral

Figure 5.3 Split wiring of duplex receptacle. Upper half is switch controlled, lower half is hot all the time.

When running overcurrent protection that is not an integral part of the disposer unit, the electrician must install separate protection; however, most disposals manufactured today have a built-in thermal protector to comply with the NEC requirements.

NEC Section 442-8(d) now permits a three-conductor cord not less than 18 inches and not over 36 inches with a grounding-type attachment plug to be used, providing the receptacle is accessible and located to avoid physical damage to the flexible cord. All of the above conditions must be met. A new exception to the rule is that if the disposal is protected by a system of double insulation or its equivalent and is marked, it shall not be required to be grounded.

Usually a 20-ampere circuit will be sufficient for a 120-volt disposer.

5-8
Bathroom Heater

Because of lack of wall space, ceiling heaters are often installed in bathrooms. Three types of ceiling heaters are available; one with a resistance element, one with a resistance heater with a motor driven fan circulator, and one with infared lamps. The resistance type heater is sometimes combined with a fluorescent lamp or with an incandescent lamp to provide light and heat at the same time (called a Heat-A-Vent light unit). See Figure 5.4.

(a)

Figure 5.4 (a) Bathroom heat-a ventlite, provides heat and ventilation together; ventilation and light together; ventilation only or light only. (b) Two lamp infrared. (c) Three-lamp infrared ceiling heater. (Courtesy of Nutone Division, Scovill Manufacturing Company)

9420

(b)

9430

(c)

Figure 5.5 Bathroom wall heater. (Martin Industries)

The infrared lamp type, which also performs both functions, could be installed with one, two, or three lamps (Figure 5.4).

All of the above could be controlled by a timer located on the wall near the light switch. By turning the timer, the heater would run the required length of time, then automatically shut itself off.

If wall space allows, a resistance 1250 or 1500 watt wall heater could be installed (Figure 5.5).

Here again the heater name-plate capacity rating dictates wire sizes. Usually an independent circuit should be provided.

The bathroom ground-fault receptacle was discussed in Chapter 4.

**5-9
Garage Door Opener Outlet
(Figure 5.6)** Automatic garage doors are designed for convenience, safety, and security. Several model variations are manufactured to accommodate different size and type doors and door hardware. An outlet must be installed overhead, centered on the door, usually 6 to 8 feet from the door opening. This outlet must now be a ground-fault circuit interrupter. NEC 210-8 states "all 120-volt single phase, 15-20 ampere receptacles installed in the garage of dwelling units shall have ground-fault." Ground-fault was discussed in Chapter 4.

**5-10
Attic Ventilator Fan
(Figure 5.7)** Intense summer heat often builds up in attics to 140° F or more. It's like a hot blanket over the entire house. An attic ventilator can reduce this attic heat to a temperature approaching that outside, permitting your air conditioning to work more efficiently, and in some homes, eliminate the need for air conditioning.

A minimum of one square foot of inlet area per every 300 CFM is required for proper fan operation. Inlet area should be located as far from the fan location as possible. More efficiency is achieved when air-intakes are installed in eaves.

How to Compute Size Needed To determine the proper size of an attic ventilator fan needed for the average home, multiply the square feet of attic area by 0.7 CFM. This will give you the minimum air delivery needed to ventilate the attic.

Example:
1500 sq ft attic area X 0.7 CFM = 1050 CFM.
For dark roofs, add 15 percent to the required CFM.
For exceptionally large attic areas or split-level homes, use two or more ventilators to provide the needed air delivery.

Figure 5.6 Automatic garage door opener. Power unit is completely enclosed in steel housing to guard mechanism against moisture and dirt. Safety features include automatic reverse cut-off, time/delay light, overload clutch, automatic stop, manual release ring, and security locks. (Courtesy of Nutone Division, Scovill Manufacturing Company)

Summary 1. Circuits for appliances may be either 120/240 volts.
2. The NEC allows a demand factor for electric ranges and other cooking appliances over 1¾ kW.
3. A separate circuit should be installed for microwave ovens.
4. The power consumed by a water heater varies from 700 watts to 5000 watts.
5. The NEC applies to the installation of most special-purpose outlets.
6. A separate circuit must be provided for a clothes dryer.
7. The dishwasher should be connected to a separate 20-ampere circuit.
8. The garbage disposer requires running overcurrent protection not to exceed 125 percent of the full-load current rating of the motor.
9. Infrared lamps could be installed for both light and heat in the bathroom.

Figure 5.7 Attic ventilation fan. (*a*) Shown installed in pitched room between 14½" rafters, can be used with standard on-off switch for manual control or with automatic thermostat. (Courtesy of Nutone Division, Scovill Manufacturing Company)

(*a*)

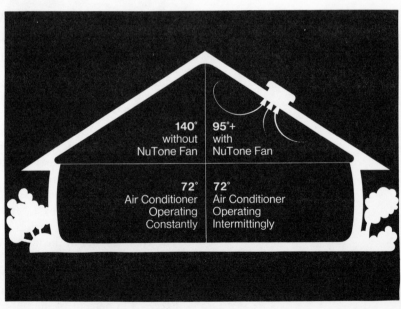

(*b*)

10. The NEC requires garage door openers to be on a ground-fault circuit.

11. An attic ventilator fan can reduce attic heat temperature, permitting the air conditioner to work more efficiently.

Problems **5-1** A 6-kW countermounted cooking unit and a 4-kW wall-mounted oven are to be installed in the kitchen of a dwelling. Calculate the maximum demand according to table 220-19 of the NEC.

5-2 A wall mounted oven is rated at 7.5 kW. How many watts is this equal to?

5-3 The NEC states that water heaters having a capacity of 120 gallons or less shall be considered continuous-duty, and as such, the circuit must have a rating of not less than what percent of the rating of the water heater?

a. 80 percent

b. 70 percent

c. 125 percent

5-4 A small window air conditioner draws 13 amperes. To what size circuit must it be connected?

5-5 A free-standing range is rated at 11.8 kW, 240 volts. According to Table 220-19 of the NEC, what is the maximum demand?

Part 3 Signal Circuits

Chapter 6

Signal Systems

Instructional Objectives

1. To provide the basic instructions for roughing-in telephone/ television outlets.
2. To make you aware of the need for residential fire/intruder alarm systems.
3. To become familiar with the location and installation of smoke detectors.
4. To learn how to plan for and lay out a radio/intercom system.
5. To become more familiar with components for residential security alarm systems.
6. To understand the operation of a two-note/two-door door chime.
7. To learn what the National Electrical Code requirements are for low-voltage conductors.

Self-Evaluation Questions

Test your prior knowledge of the information in this chapter by answering the following questions. Watch for the answers as you read the chapter. Your final evaluation of whether you understand the material is measured by your ability to answer these questions. When you have completed the chapter, return to this section and answer the questions again.

1. What requirements cover the installation of telephone outlets?
2. What must the electrician provide in the conduit run for the telephone installer?
3. Who should be consulted before work is started on television outlets?
4. Is it necessary to nail up an outlet for all television outlets?
5. Explain how the tone is made in a two note/two door door chime.
6. How many solenoids are contained in a two-note chime?
7. What is a signal circuit?
8. Where are smoke detectors located?
9. What is the advantage of installing a fire/intruder alarm system?
10. What are the NEC requirements for low-voltage signal circuits?

6-1
Signal Equipment

Every residence is equipped with some signal equipment. The most simply constructed house has a doorbell and usually a telephone. A more expensive, modern, custom-built home will most likely have prewired telephone jacks, amplified television antenna outlets, a radio/intercom system, and a fire/intruder alarm system.

In Chapter 1, Article 100 — Definitions, The National Electrical Code defines a signaling circuit as "any electric circuit that energizes signaling equipment."

6-2
Telephones

Usually the installation of the residential telephone system is done according to the requirements of the local telephone company. In some areas the telephone company will have its own construction crews to rough-in a house; that is, the company furnishes and installs outlet boxes or plaster rings, cables, phone jacks, wall plates, connecting blocks, and other materials necessary to complete the installation.

In some areas, the electrician will rough-in the telephone outlets by nailing a raised plaster ring without the outlet box directly to the 2 X 4 stud at the same height as a duplex receptacle. A fish-wire is dropped from each outlet to the basement. When the telephone company installer is ready, cables are attached to the fish-wire, which is then withdrawn from the plaster ring outlet where the final connections are made. Some difficulty is to be expected in this procedure, since insulation in the walls tends to obstruct the fish-wire.

An alternate method would be to install the fish-wire in a run of ½" thin wall conduit from each phone outlet location to an accessible point in the basement. The installer is then able to pull the cables through the conduit and terminate them at the proper location (Figure 6.1).

Most telephone companies require at least one phone to be permanently installed. The remaining phones may be of the portable extension type that can be plugged into any of the phone jacks furnished by the telephone company.

6-3
Television

There are times when the electrician is called upon to install the television antenna and run the cable to the outlets. However, television is a highly technical field. It is recommended that a compe-

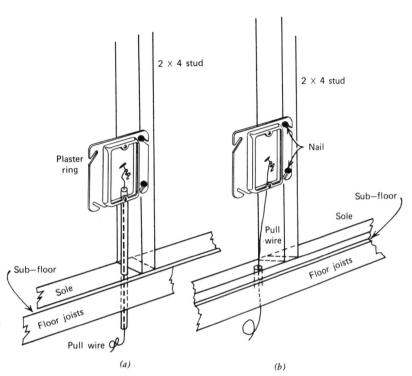

2 × 4 stud

2 × 4 stud

Nail

Plaster ring

Sub—floor

Sole

Pull wire

Sub—floor

Floor joists

Sole

Floor joists

Pull wire

Figure 6.1 (*a*) Installation with pull wire run in ½″ conduit for telephone installer. (*b*) Pull wires to accessible area in basement or attic.

(a)   *(b)*

tent television installer be consulted before any work is started on the installation.

There are a number of ways in which television outlets may be installed. In general, a single-gang plaster ring may be used at each point on the system where an outlet is located. Nail the plaster ring to a 2 X 4 stud at the same height as the wall outlets and pull enough cable in so that a connection may be made to the TV plates on the finish.

Both indoor and outdoor antennas are available for black and white and color sets. Antennas may be installed in the attic or on the roof.

Figure 6.2 shows TV outlet on stud by outlet box.

6-4
Door Chimes Most modern residential occupancies use chimes to announce the presence of someone at the door. Chimes are available in many different designs and styles. The simple two-note types are provided with two solenoids and two iron plungers. When one solenoid is

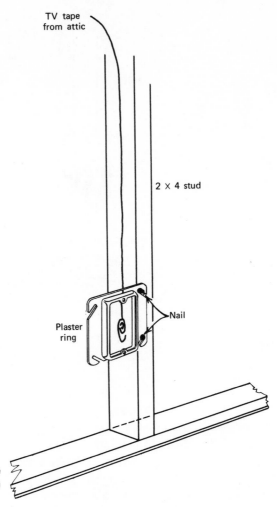

TV tape
from attic

2 X 4 stud

Nail

Plaster
ring

Figure 6.2 Television outlet complete
with pull wire or TV tape pulled in from
attic.

energized, the iron plunger is drawn into the opening of the solenoid where a peg in the end of the plunger strikes one chime tone
bar (usually the back door). When the solenoid is de-energized, the
plunger returns by spring action and comes to rest against a soft felt
pad so that it is prevented from striking the other chime tone bar.
As a result, a single chime tone is heard. When the second solenoid is energized, one chime tone bar is struck. When this solenoid
is de-energized, the plunger returns by spring action and strikes the
second tone bar. In this manner a two-tone signal is produced
(usually for front door). The plunger then comes to rest between
the two tone boxes.

Other types and styles are available in up to eight-note and repeater-tone styles. In the repeater-tone model, both notes continue to sound as long as the pushbutton is depressed.

Electronic chimes that relay their chime tones through various speakers of an intercom system are also available.

Wiring The Chime Installation

Regardless of the type of style of the chime, the instructions of the manufacturer must be followed when installing the unit.

The wire used for the chime circuit is low-voltage wire, commonly called bell wire or thermostat wire. Most thermostat wire used in today's construction is insulated with a thermo-plastic compound identified as type "T." Because of the low voltage involved and the small current requirement, No. 18 AWG conductors are usually used.

Color-coded multicolor cables of two or three single wires are contained within a single protective covering, which may be fastened by insulated staples directly to the wall studs or run along the sides of floor joists in the basement.

When roughing-in for a chime, it is a good idea to install some backing between the wall studs at the required height. Should the chime be eight-note (four tube) and heavy, the backing will give support when installed.

National Electrical Code Requirements

Article 725, Sections 31-42.

1. A general rule is: low-voltage wire with low-voltage insulation must not be installed in the same enclosure with higher-voltage light and power conductors.
2. Low-voltage wire must not be installed closer than two inches to open light and power conductors unless the low-voltage wires are permanently separated by some approved type of insulation.
3. Low-voltage wire may not enter an outlet box containing a higher-voltage light and power conductors unless the two voltages are separated by a metal barrier.

Figure 6.3 illustrates wiring diagrams for door chimes (Nutone).

6-5 Radio-Intercom Systems

Combination units such as that in Figure 6.4 have become very popular. The system not only consists of a master station and a number of remote stations in various rooms but is also equipped

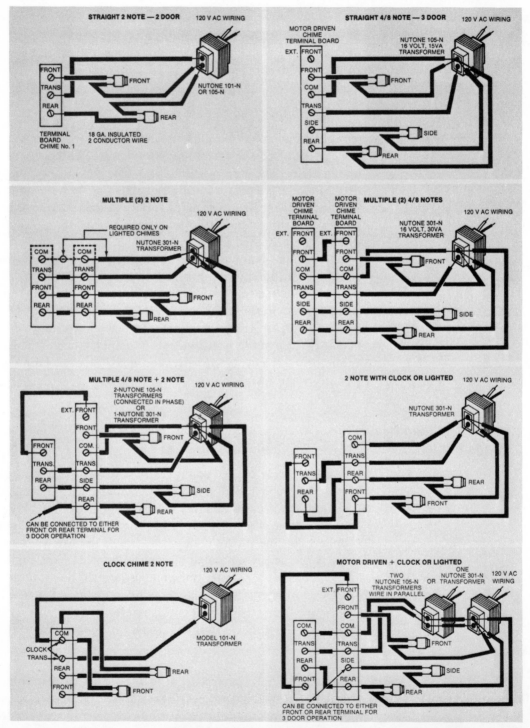

Figure 6.3 Wiring diagrams for door chimes. (Courtesy of Nutone Division, Scovill Manufacturing Company)

(a)

(b)

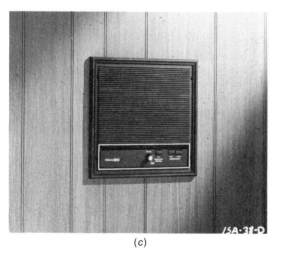

(c)

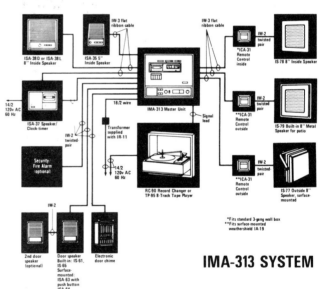

IW-3 flat ribbon cable

ISA-38 D or ISA-38 L 8" Inside Speaker

ISA-35 5" Inside Speaker

*ICA-31 Remote Control inside

IW-2 twisted pair

IS-78 8" Inside Speaker

14/2 120v AC 60 Hz

ISA-37 Speaker/ Clock-timer

18/2 wire

IMA-313 Master Unit

Signal lead

IW-2 twisted-pair

Transformer supplied with IR-11

**ICA-31 Remote Control outside

IS-79 Built-in 8" Metal Speaker for patio

Security Fire Alarm (optional)

14/2 120v AC 60 Hz

RC-90 Record Changer or TP-95 8-Track Tape Player

**ICA-31 Remote Control outside

IW-2 twisted-pair

IS-77 Outside 8" Speaker, surface-mounted

*Fits standard 3-gang wall box
**Fits surface-mounted weathershield IA-19

IW-2

2nd door speaker (optional)

Door speaker Built-in: IS-61, IS-65 Surface-mounted: ISA-63 with push button ISA-64

Electronic door chime

IMA-313 SYSTEM

(d)

Figure 6.4 Radio intercom communicenter. (a) Radio intercom equipped with a cassette tape player/recorder as a family message center. Accommodates up to 12 remote stations. (b) Surface mounted door speaker with push button control. (c) Eight-inch inside speaker. (d) Wiring diagram. (Courtesy of Nutone Division, Scovill Manufacturing Company)

with a cassette tape player/recorder as a family message center. This centralized system accommodates up to 12 remote stations. Even door speakers have "hands-free" answering. Calls originating from remote stations are answered by pressing the "talk" button. Auxiliary input jacks allow addition of a foldaway record changer and 8-track tape player.

Connection of these devices by the electrician is relatively simple because the manufacturer furnishes color-coded wire, numbered terminal blocks, and detailed installation instructions.

6-6
Residential Intruder —
Fire Alarm Systems

There are several varieties of dependable electronic protection devices on the market to give early warning of fire, lethal smoke, or forced entry. They are designed to protect the home and family from fire and intruders. Most of these are sensing devices — they sense through sound and motion and announce the presence of fire or of an intruder. Some electronic systems signal an intruder's entry by alarms, by switching on lights, or by direct phone lines to the police station.

Figure 6.5 shows a fire/intruder alarm system designed for residential use. It has a master control unit, a series of intruder and fire detectors, and indoor/outdoor alarms. The control unit incorporates its own alarm horn, optional battery stand-by, and easy-to-use controls. Modular in design, the unit is solid-state with fuse protection circuitry. The control monitors heat and smoke detectors and supervises a comprehensive variety of fail-safe features, component options, and alarm accessories. It is customized to fit the size of the home and the requirements of the family.

Figure 6.6 shows the master control unit.

Figure 6.7 shows fire/intruder alarm system components.

6-7
Smoke Detectors

A fire is unpredictable. A well-planned detection system should include both smoke and heat detectors in critical areas and in numbers sufficient to assure full coverage of the home.

There are two types of smoke detectors commonly used in homes: ionization detectors and photoelectric detectors.

Ionization detectors use a radioactive source to transform the air inside them into a conductor of electric current. A small current passes through this "ionized" air. When smoke particles enter the detector, they impede the flow of current. Electronic circuitry mon-

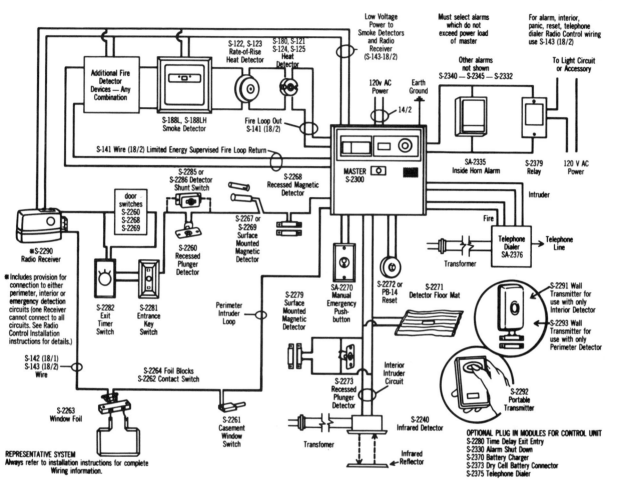

Figure 6.5 Intruder-fire alarm wiring diagram. (Courtesy of Nutone Division, Scovill Manufacturing Company)

itors the current reduction and sets off an alarm when the current gets too low.

Photoelectric detectors have a lamp that directs a light beam into a chamber. The chamber contains a light-sensitive photocell, which is normally tucked out of the way of the lamp's direct beam. But when smoke enters the chamber, the smoke particles scatter the light beam. The photocell now "sees" the light and, at a preset point, sets off an alarm.

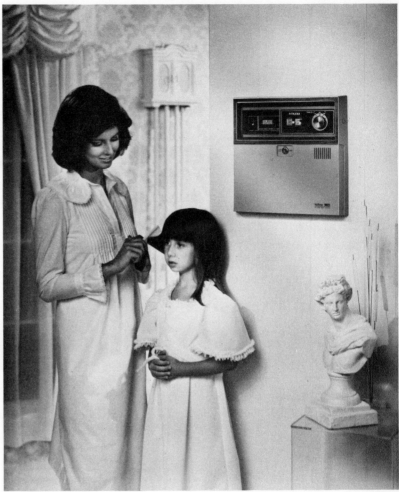

Figure 6.6 Intruder-fire alarm master control. (Courtesy of Nutone Division, Scovill Manufacturing Company)

For maximum protection: install a smoke detector (Figure 6.8) outside each bedroom area and at the top of stair wells; and heat detectors (Figure 6.9) in each enclosed living area, including bathrooms, closets, attic, and basement.

It is recommended that at least one smoke detector be installed in the living level of multi-story homes.

Install smoke alarms on the ceiling in the center of the selected area or mount on the wall with the top of the alarm not less than 6 inches and not more than 12 inches from the ceiling/wall junction.

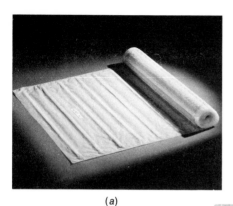

(a)

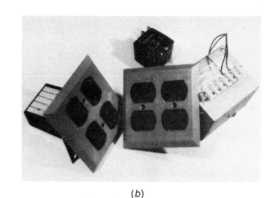

(b)

Figure 6.7 Intruder-fire alarm system components (a) Detector floor mat for easy concealment under rugs at doors, windows, or stairs. (b) Infrared face plate detector, protects entries, hallways, rooms, any large area; covers a span from 3 to 75 feet. (c) Outside electronic siren alarm. (d) Heat detector. It activates alarm when present temperature limit is exceeded, causing the contacts to close. (Courtesy of Nutone Division, Scovill Manufacturing Company)

(c)

(d)

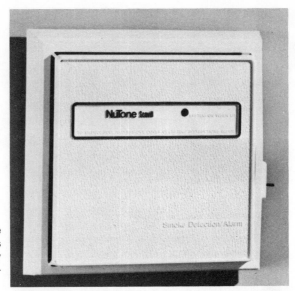

Figure 6.8 Smoke detectors. Surface mounted on wall or ceiling in hallways or other access to bedrooms. (Courtesy of Nutone Division, Scovill Manufacturing Company)

Figure 6.9 Heat Detector. Rate-of-rise 135° fixed temperature. Surface mounted on ceiling or in ordinary living areas with normal room temperatures. (Courtesy of Nutone Division, Scovill Manufacturing Company)

Summary
1. Telephone company requirements must be followed when installing telephone outlets.
2. When roughing-in a television outlet, a single-gang plaster ring is usually nailed to the wall stud.
3. Low-voltage door chime installations are covered by rules of the National Electrical Code.
4. A radio/intercom system consists of a master station and a number of remote stations in various rooms.
5. For convenience the radio/intercom master station is usually installed in the kitchen area.
6. Residential intruder and fire alarm installations are dependable protection devices.
7. For protection, smoke detectors should be installed in or near each bedroom.

Problems
6-1 According to the National Electrical Code's definition, how would you define a signaling circuit?
6-2 Make a list of the signaling outlets and equipment usually installed in a more expensive custom-built home.
6-3 Explain the procedure for roughing-in a telephone outlet.
6-4 Draw a wiring diagram for a 2-note/2-door door chime; show transformer, front-rear door buttons, and terminal board of one chime.
6-5 Draw a simple sketch of a hallway wall. Show where you would install a smoke detector for maximum protection.
6-6 Figure 6-5 shows a wiring diagram of an intruder/fire alarm system. From this diagram, locate and list at least six components to indicate an intruder setting off the alarm.

Part 4 Heating Systems

Chapter 7

Electric Space Heating

Instructional Objectives

1. To learn the advantages of installing electric space heating in a residence.
2. To become more familiar with the methods of heating.
3. To understand the different types of electric heating systems.
4. To learn how to install electric space heating with appropriate temperature control.
5. To become familiar with the difference between baseboard and ceiling heating cables and the installation of each.
6. To learn how to calculate heat loss.
7. To learn the requirements of the NEC for the installation of electric space heating.

Self-Evaluation Questions

Test your prior knowledge of the information of this chapter by answering the following questions. Watch for the answers as you read the chapter. Your final evaluation of whether you understand the material is measured by your ability to answer the questions. When you have completed the chapter, return to this section and answer the questions again.

1. What are the advantages of installing electric space heating units?
2. List the different types of electric heating systems.
3. What are the three methods of heating?
4. Normally, on which walls should thermostats be located?
5. What is the recommended elevation above the floor that thermostats should be located?
6. How is a heating cable installed in the ceiling of a residence?
7. According to the NEC, what code limitations exist for line thermostat controllers for fixed electric space heating?
8. What code requirements exist for splicing heating cables?

9. Why is closer spacing of radiant heating cables recommended near outer walls and windows?

10. What is heat loss?

7-1 Electric Heating Unlike combustion heating, electric heating leaves no residue. "Pure" energy in the form of electricity flows into the house through wires. There is no waste; all of the energy expended is converted into heat, and this is why we call electricity "pure" energy. Combustion of fuel may take place, but the fire is far away in the boiler of the power plant.

Heating an entire house requires a lot of heat, many thousand BTU's per hour, regardless of the method of heating. Electrically, the load during the heating season may be continuously 10 to 20 kilowatts. This is not an unusual or especially large load, but when contrasted with that of a typical home having only lighting and normal household appliances, there is quite a difference.

This means several things. To homeowners just becoming acquainted with electric heating, their electric bills are higher than they are used to. More of the cost of the home is in the electrical system. And, there may be other economic considerations, but these have to be evaluated in terms of the benefits. However, they will not be discussed here.

To the electrical contractor, it means checking the available electrical service carefully before selling the homeowner a job; knowing additional building codes and regulations; and installing bigger, more complex, heavier electrical services as well as more equipment and controls.

To the utility company, it means having to provide completely satisfactory electrical service. To do this may involve such changes as a considerable increase in generating capacity, new regulations allowing higher voltages, and different electrical rates.

Surely, more people need to know more about the use of electricity in heating installations.

7-2 Electric Heat In Perspective In the past year, electrical usage has soared as never before. Everywhere we look, electrical energy is being used in more and bigger ways — light, heat, and power. It now seems clear that in the late 1970's electrical energy will assert itself as clearly superior to all other forms of energy both now and in future generations.

It is interesting to note that the shocking world-wide energy crisis — with its oil and gas shortages hitting all segments of the economy — has been a major stimulus of the growth of electrical application. Even though the push to nuclear generation of electricity has been slowed down by an assortment of technical, ecological, and economic problems, it will surely be one of the dominant power sources of the future. Another dominant source will be the harnessing of the sun's radiant waves — the use of "solar energy." The pace of developing these power sources that lead to the total electrical energy age will be even faster than anyone might estimate, since research, development, economics, and ecology are all on the side of electricity.

The growing availability of electricity, its ease of installation, its flexibility and cleanliness at the point of use, and many other factors, coupled with the steadily dwindling availability of fossil fuels, have all played significant roles in causing the remarkable growth of electrical energy usage, especially in the areas of electric space heating and comfort conditioning.

When economic conditions are favorable, heating by electricity is found to have many advantages.

7-3 Advantages

1. *Flexible:* Just as electricity is flexible in terms of generation methods, so is electric heat flexible at its point of use. It can be used for completed heating or just partial heating. For example, an electric heating unit can be installed to supply heat to just one room—such as an addition to a house—without affecting whatever other system may be involved. Electric heating units can be used to heat just one area without having to start and operate the main heating system. Unlike other heating systems, electric heating units are available for almost all types of applications, including single and multifamily housing units.

2. *Clean:* The process of converting electricity into heat involves: no dirt, no dust, no vapors, no fumes or odors. Because electric heating units typically include equipment that limits the amount and velocity of air, there is less likely a discoloration of walls; and because electrically heated homes usually involve better insulation and weatherproofing, there is less chance of infiltration of dust and dirt from outside air. In fact, electricity is the cleanest source of energy available for producing heat.

3. *Efficient:* Electric heat is 100 percent efficient so far as conver-

sion of electric energy into heat is concerned. It requires no air for combustion; therefore there is no heat loss up a chimney.

4. *Durable:* In many of today's homes, fuel-fired systems will require eventual replacement due to overall deterioration caused by the combustion process. Electric heating units can be expected to last the life of the heating element, usually comparable in cost to a single cleaning of a fuel-fired system.

5. *Quiet:* Electric heating units — most of those now being manufactured — are silent. This can prove very valuable in helping to control noise pollution, especially when a well-insulated and weatherproofed dwelling is involved.

6. *Controllable:* Unitary electric heating equipment usually has a thermostat of its own. As a result, a homeowner can adjust the temperature of the room to meet the requirements of comfort for that area. On the other hand, single-thermostat installations require that the system provide heating for the entire space served by the system, which can be wasteful unless motorized dampers and other expensive, intricate mechanical controls are utilized. In fact, sophisticated new solid-state controls for electric heating units are being developed, resulting in more selective and energy-efficient use of heat.

7. *Completely comfortable:* There are no drafts, no cold spots, no uneven irregular heat.

8. *Good environmental impact:* Electric space heating emits no air pollution at the home site. The pollution control at the utilities' generating plants are so effectively controlled that pollution there is only a small percentage of what it would be if fossil fuel were burned at each home.

7-4
Heat Transfer

It must be understood that heat is a form of energy and therefore cannot be created or destroyed. However, it can be moved or transported from one place to another through varied mediums.

In order to understand how a heating system works, it is necessary to understand the ways in which heat transfer can occur. Water always flows downhill, never uphill, always from a higher level to a lower level. In a similar way, one might think of heat as always flowing in one direction, from a position of higher temperature to one of lower temperature.

When water flows downhill, the steeper the hill the faster the water travels. Likewise in the transfer of heat, the greater the temperature difference, the greater the quanity of heat will flow in a unit of time. There are three main ways the transfer of heat takes place: conduction, radiation, and convection.

Conduction
This is the flow of heat through a substance due to the transfer of heat energy from particle to particle, from a warmer region to a colder region. For example, if a rod is heated over an open flame, heat travels by conduction from the hot end to the cooler end. Conduction heat transfer occurs not only within an object or substance but also between different substances that are in contact with one another.

A good example of this is house construction. In house construction, there is a combination of wood, insulation, sheetrock, plaster, brick, and even concrete. These materials are often touching each other. If it's hotter in the house than outdoors, heat by conduction will pass through these materials. If it's hotter outdoors than it is in the house, heat will flow into the house (Figure 7.1).

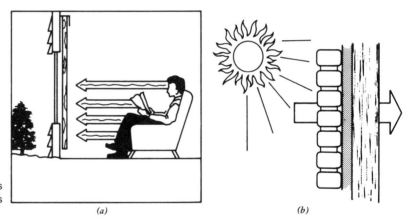

Figure 7.1 Conduction. (*a*) Heat flows from hot to cold. (*b*) Conduction takes place between different materials.

(*a*)   (*b*)

Radiation
Radiation is the transfer of heat through space just as light travels through space. Radiant heat passing through the air does not warm the air through which it travels. Radiant energy travels in straight lines until it is interrupted or absorbed by some object or body. When this energy strikes an object, it causes the molecules of the absorbing object to vibrate. This vibration coverts the energy into heat. Radiant energy will pass through certain materials such as

Figure 7.2 Radiant energy travels in straight lines until it is interrupted or absorbed by some object or body. (Courtesy of The Singer Company, Climate Control Division)

glass without heating them, yet it is reflected by various other materials. It will pass through air regardless of its temperature without heating it to any appreciable extent (Figure 7.2).

A good example of the effect of radiant heat is when you are lying on the beach with your face in the sand. You are being warmed by energy waves from the sun. Suddenly, you feel cool. You know immediately, without turning to look, that a cloud has come between you and the sun. The air temperature remains the same, yet you feel cool because your source of "radiant heat" has been blocked.

Convection Convection heating is where the natural upward flow of heated air results from contact with convective surfaces that have been heated by an electric current (Figure 7.3).

7-5
Types of Heating Units There are many styles and designs of electric space heating units manufactured in this space age. However, the type unit to be installed will depend upon several factors: economic conditions,

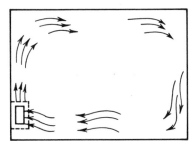

Figure 7.3 Convection heating. Heat transfer depends on air movement over the heated surface in the heating equipment and then into the room. The circulation of air over the convector surface may be produced by a fan or blower or by natural convection.

structural conditions, the kind of room and the purpose for which the room is to be used, and whether the homeowner should plan for the future installation of a "heat pump" or for converting the unit to "solar energy". All of the above will be considered as we look at the different types.

Baseboard Heaters The baseboard heater is used for general comfort heating of a residence. Baseboard sections are available in a wide range of heat output ratings. However, standard density heaters (250 watts per foot) are the most commonly used. They may be used in single sections or joined to supply the heat needed for any given area. These sections can be employed as a sole heating source or as an addition to a central plant, heat pump, or solar energy unit. This type of heater is installed at the floor line, usually on an outside wall under the windows (Figure 7.4).

Baseboard heaters have the advantage of taking a minimum of space in the room and of delivering the heat along a broad area at the outer perimeter of the room where it is needed. In this location, the cold air will be warmed before it circulates through the room. However, the heater should be located so that an adequate circulation of warm air will be provided to the area to be heated.

With convectors, room air flows in through an opening at or near the bottom. The air then flows around the heating elements, out through the opening at the top, and around the room.

The baseboard heater in any of its forms is suited to control by a wall mounted thermostat, though some are manufactured with built-in thermostats. Many now use a high-limit safety control along the full length of the heater to provide against hot spots in case of blocked circulation (Figure 7.5).

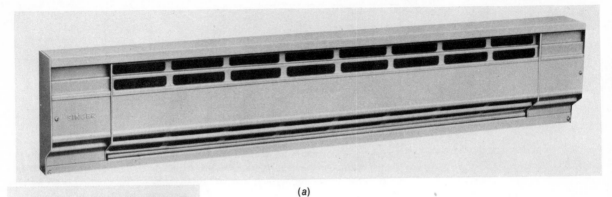

(a)

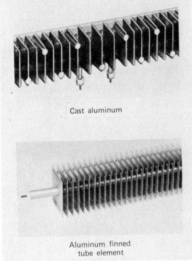

Cast aluminum

Aluminum finned
tube element

(b)

Figure 7.4 (*a*) Baseboard heating unit. (*b*) Metal heating unit elements through which the resistance wiring runs. Cast-aluminum grid retains warmth somewhat longer than the aluminum finned-tube element. (*c*) Cutaway view of a heater employing a cast-aluminum grid. End sections with removable covers may be used for wiring convenience outlets or for built-in thermostat to control the baseboard. (Courtesy of The Singer Company, Climate Control Division)

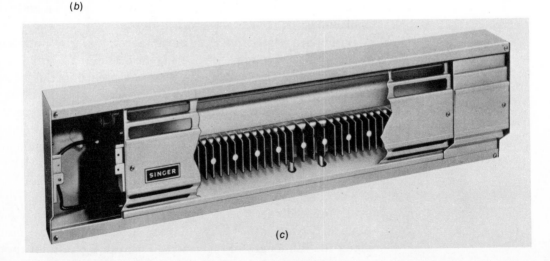

(c)

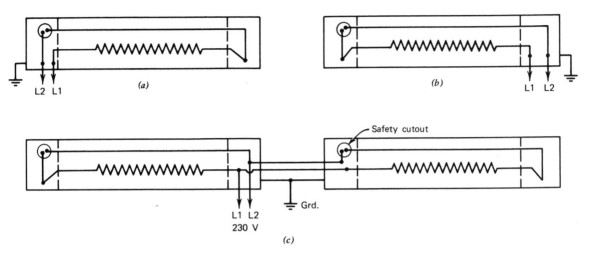

Figure 7.5 Wiring diagram for standard baseboard electric heaters. (*a*) Left-hand line connection. (*b*) Right-hand line connection with safety cutout. (*c*) Power supply to two baseboard heaters each having its own safety cutout.

Cove Heater Systems This is a radiant heating unit installed on the wall near the ceiling as a valance. The unit works on the principle that uniform heat emissions of the radiant type will blanket the entire room, directly heating objects in the room (Figure 7.6).

It is the same kind of heat that warms you instantly when the sun strikes a window pane. You are warm as long as the sun is shining and cooler when the sun no longer shines on the window.

Cove units usually come in standard lengths of 42, 60, and 84 inches to accommodate the proper wattage for heating in any room. See Figure 7.7 for ease of installation.

Units are thermostatically controlled so a homeowner will get zoned heat in each room. Temperature requirements are met quickly, quietly, and silently without chilly drafts, blasts of hot, dry air, soot, or dust. Since there are no moving parts to the heater, it is fool-proof and maintenance free (Figure 7.8).

Electric Furnace Electric furnaces are basically similar to fossil fuel furnaces except that electric resistance heating elements replace the burner (Figure 7.9). Generally, a number of electric heating elements are used. The furance cabinet also encloses the air filter and a circulating fan that forces the air through the heating section and ductwork to the various rooms. Temperature-limit controls prevent the furnace's overheating. Comfort control of the system is provided by a low-voltage thermostat and a relay or sequencer and fan control.

ELEKTRA-COVE heating effectively and efficiently gives instant satisfying comfort to people of all ages...providing a gentle warmth that permeates every corner of the room...with a uniform heat distribution from floor to ceiling unmatched by any other single system. It is accomplished in a most unique manner, through Primary and Induced radiant wave energy beamed into the space being conditioned.

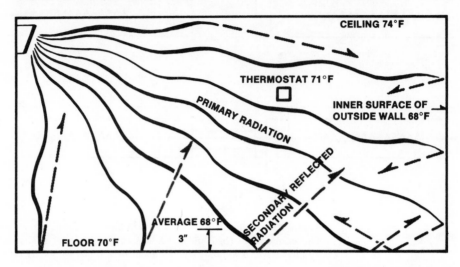

Figure 1 shows Primary Radiation of wave energy being emitted from the face of the unit. This wave energy has the property of directly heating all walls, objects and people. Depending on the type of surface it strikes a percentage of wave energy is absorbed and the balance is reflected and again comes in contact with other surfaces. The energy which is reflected is referred to as secondary radiation and it is this energy, coming from all directions that contributes to uniform comfort heating.

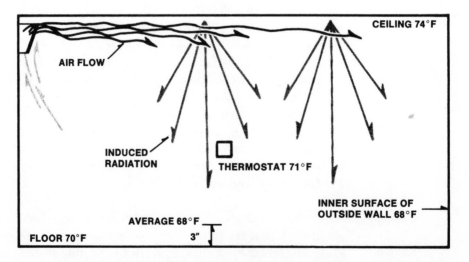

Figure 2 shows how induced radiation is caused. Since the ceiling has a considerable mass it acts as a heat bank which holds the heat and thus radiation continues to distribute warmth throughout the entire room even after the thermostat is satisfied, resulting in more uniform temperatures.

Because of its design features the **Elektra-Cove** enables the induced heat from the ceiling together with the highly efficient (75 to 80%) direct radiation from the face plate to provide satisfying comfort within moments after the system is turned on.

Figure 7.6 Cove radiant heating. (Courtesy of Elektra Systems, Inc.)

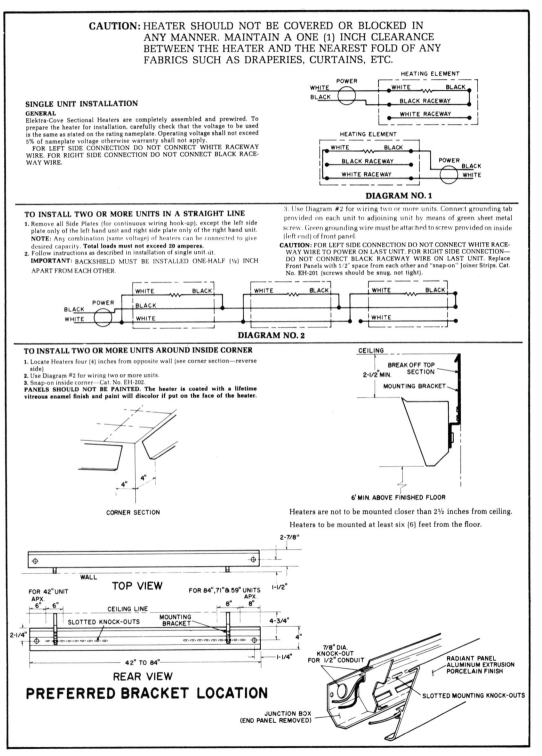

CAUTION: HEATER SHOULD NOT BE COVERED OR BLOCKED IN ANY MANNER. MAINTAIN A ONE (1) INCH CLEARANCE BETWEEN THE HEATER AND THE NEAREST FOLD OF ANY FABRICS SUCH AS DRAPERIES, CURTAINS, ETC.

SINGLE UNIT INSTALLATION

GENERAL

Elektra-Cove Sectional Heaters are completely assembled and prewired. To prepare the heater for installation, carefully check that the voltage to be used is the same as stated on the rating nameplate. Operating voltage shall not exceed 5% of nameplate voltage otherwise warranty shall not apply.

FOR LEFT SIDE CONNECTION DO NOT CONNECT WHITE RACEWAY WIRE. FOR RIGHT SIDE CONNECTION DO NOT CONNECT BLACK RACEWAY WIRE.

DIAGRAM NO. 1

TO INSTALL TWO OR MORE UNITS IN A STRAIGHT LINE

1. Remove all Side Plates (for continuous wiring hook-up), except the left side plate only of the left hand unit and right side plate only of the right hand unit. **NOTE:** Any combination (same voltage) of heaters can be connected to give desired capacity. **Total loads must not exceed 20 amperes.**
2. Follow instructions as described in installation of single unit.

IMPORTANT: BACKSHIELD MUST BE INSTALLED ONE-HALF (½) INCH APART FROM EACH OTHER.

3. Use Diagram #2 for wiring two or more units. Connect grounding tab provided on each unit to adjoining unit by means of green sheet metal screw. Green grounding wire must be attached to screw provided on inside (left end) of front panel.

CAUTION: FOR LEFT SIDE CONNECTION DO NOT CONNECT WHITE RACE-WAY WIRE TO POWER ON LAST UNIT. FOR RIGHT SIDE CONNECTION— DO NOT CONNECT BLACK RACEWAY WIRE ON LAST UNIT. Replace Front Panels with 1/2" space from each other and "snap-on" Joiner Strips. Cat. No. EH-201 (screws should be snug, not tight).

DIAGRAM NO. 2

TO INSTALL TWO OR MORE UNITS AROUND INSIDE CORNER

1. Locate Heaters four (4) inches from opposite wall (see corner section—reverse side)
2. Use Diagram #2 for wiring two or more units.
3. Snap-on inside corner—Cat. No. EH-202.

PANELS SHOULD NOT BE PAINTED. The heater is coated with a lifetime vitreous enamel finish and paint will discolor if put on the face of the heater.

CORNER SECTION

Heaters are not to be mounted closer than 2½ inches from ceiling.

Heaters to be mounted at least six (6) feet from the floor.

TOP VIEW

REAR VIEW
PREFERRED BRACKET LOCATION

Figure 7.7 Wiring diagram for cove radiant heating. (Courtesy of Elektra Systems, Inc.)

Figure 7.8 Cove heating units installed in a kitchen. (Courtesy of Elektra Systems, Inc.)

The electric furnace, with a properly designed duct system, may be combined readily with central cooling systems and an electronic air cleaner to provide year-round air conditioning.

Central Fan With Duct or Register Heaters Warm-air duct or register-heater systems do not employ a large central furnace as the major heat source. They include a central circulating fan or blower and air cleaning system with trunk and lateral ducts to each room. The main heat supply for each room is from a heater in the duct (Figure 7.10) or in the diffusing register in the room.

The central fan system offers advantages of individual room or zone control and is adaptable to central year-round air conditioning.

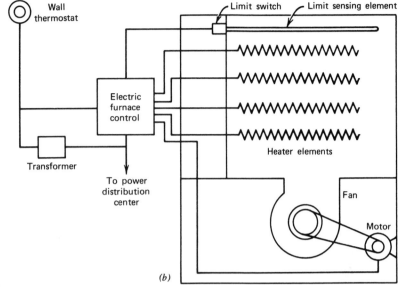

(a)

Figure 7.9 (a) Warm air electric furnace. (b) Wiring diagram showing controls and heating elements. (Courtesy of Martin Industries)

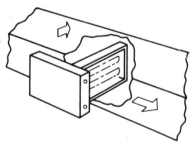

Figure 7.10 Duct heater. Basic unit is complete with heater frame, heating elements, control compartment with hinged or removable cover, primary silent controls, back-up contactors, automatic reset thermal cutout, manual reset thermal cutout, fan interlock, fan motor terminal board, factory wired.

A duct heater may supply one room or several rooms and be controlled by a zone thermostat. Register heaters installed in each room are controlled by individual room thermostats.

Direct and register units can be used to augment a "heat pump" or "solar energy" system in extreme weather. They may be used similarly to supplement any ducted heating plant having hard-to-heat rooms or additions.

Radiant Heating Panels Fabricated panels rely on heat-producing wires embedded in plasterboard, tempered glass, or aluminum circuitry sandwiched between two flat sheets of durable polyester to provide radiant heat. These panels are ideal for entryways, vestibules, and other places where a concentrated heat is needed.

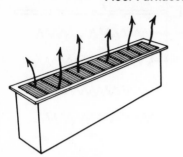

Floor Furnaces These are often used under large picture windows where there is no space for baseboard units and for numerous other situations where the floor location of the heater is preferred. They are designed to heat by gravity convection of warm air from the heater into the room. They may be made with a built-in thermostat; however, they perform better if controlled by a wall-mounted room thermostat (Figure 7.11).

Figure 7.11 Floor furnace.

Radiant Wall Heaters Radiant wall heaters offer advantages in bathrooms, entryways, and other areas where warmth is desired. They should be installed so that the heat rays will not be blocked from the desired heating target or area. They must "see" the heating target in order to heat it (Figure 7.12).

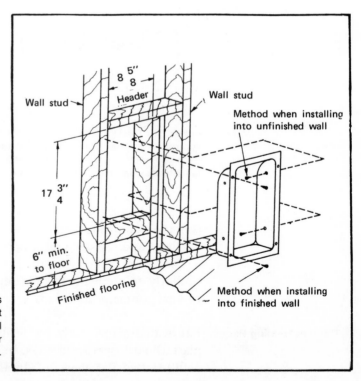

Figure 7.12 Radiant wall heater offers advantages of instant concentrated heat which makes them particularly useful in bathrooms, entryways, and in other areas where quick warmth is desired. (Courtesy of Martin Industries)

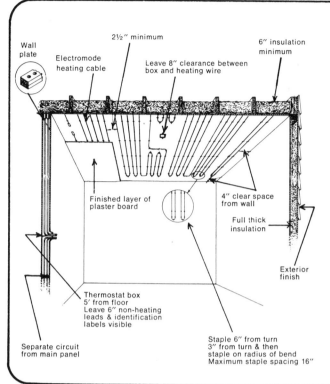

Wall plate

Electromode heating cable

2½" minimum

Leave 8" clearance between box and heating wire

6" insulation minimum

Finished layer of plaster board

4" clear space from wall

Full thick insulation

Exterior finish

Thermostat box 5' from floor
Leave 6" non-heating leads & identification labels visible

Separate circuit from main panel

Staple 6" from turn
3" from turn & then staple on radius of bend
Maximum staple spacing 16"

installation information

The diagram above depicts a typical installation for a dry wall ceiling. The cable is stapled directly to the ceiling lath, which should be of a non-metallic fire resistant type such as gypsum board or sheet rock. Under no circumstances should the cables be shortened. In the event that the factory labels have been inadvertently removed from the spool, the cable can be identified by referring to the identification tag on the non-heating leads, or by connecting suitable test meters into the circuit.

Full details on installation procedures can be found in the instruction manual which accompanies the cable.

Figure 7.13 Installation of electric radiant heating cables. (Courtesy of The Singer Company, Climate Control Division)

Electric Radiant Heating Cables For complete invisible heating, the cable can be embedded in plaster or between two sheets of wallboard. Installation of the cable is permanent and care should be exercised to avoid any abnormal condition that would cause damage to the cable or circuit overload. Ceiling electric heating cable radiates heat evenly to rooms below (Figure 7.13). Proper installation is of utmost importance in deriving full efficiency from the system.

Each room has temperature control permitting room-to-room variations for individual needs.

Wall-Mounted Convectors Wall-mounted warm air convectors offer the advantage of room air circulation that distributes warmth throughout the space. Fan or blower models produce more positive circulation and more rapid distribution of heat (Figure 7.14). For best results, wall convectors should be installed where circulation of air to the room will not be blocked.

Wall heaters are usually equipped with either wall-mounted or cabinet-mounted thermostats.

Usually approved listed heaters are equipped with a limit switch to shut down the heater and prevent overheating if the air flow should be stopped for any reason.

Figure 7.14 Wall-mounted convectors offer the advantage of room air circulation which distributes warmth throughout the space. Fan or blower models produce more positive circulation and more rapid distribution of heat (Courtesy of Martin Industries)

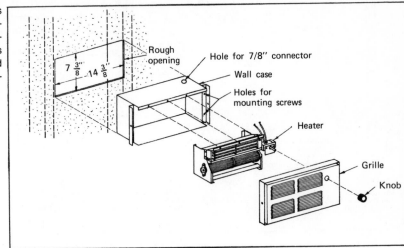

Ceiling Heaters Ceiling heaters are often used in bathrooms. In general, two types are available: one having a resistance element, the other infrared lamps. The resistance type is usually combined with a circline fluorescent lamp or with an incandescent lamp to provide light and heat at the same time. The infrared unit, which also performs both functions, utilizes one, two, or three infrared lamps (Figure 7.15).

7-6
Temperature Control
Thermostats

Years ago homeowners had only one heat control—themselves. When it was cold, they turned up the heat. When the house was too hot, they turned it down. Some of the first thermostats were crudely constructed and allowed temperature variations of several degrees. However, today the home owner has a wide choice of controls to fit their exacting heating requirements.

This section is aimed at covering some of the basic types of controls for electric space heating.

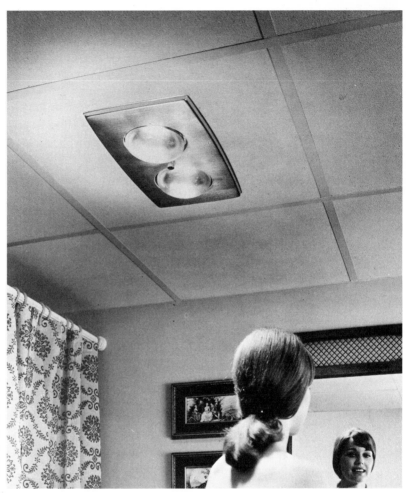

Figure 7.15 Two-bulb heat-a-lamp. Provides luxurious auxiliary heat in bathrooms. May be wired for one or two bulb operation. Use 250W R40 infrared heat resistant glass lamps. Quiet cooling system regulates internal temperatures, has automatic reset thermal protector. (Courtesy of Nutone Division, Scovill Manufacturing Company)

1. Single-Pole Line-Voltage Thermostat (Figure 7.16).
   It has one set of contacts that act as an electric switch. They are connected in the service line, in series with the heating load. Since the single-pole thermostat breaks only one side of the line, it isn't considered to have an electrical "off" position.
2. Double-Pole Line-Voltage Thermostat (Figure 7.17).
   It has two sets of contacts connected to the circuit so that one set breaks one side of the line and the second set breaks the other side of the line. Since both sides are broken when the thermostat contacts are locked in the "open" position, the double-pole unit should be used when it is desirable to break both sides of the line and when codes require such a break.

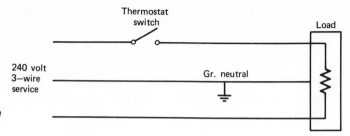

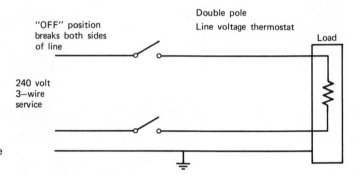

Figure 7.16 Single-pole line voltage thermostat.

Figure 7.17 Double-pole line voltage thermostat.

Thermostats built into electric heating equipment are, by their nature, line voltage. Built-in thermostats are either fixed or optional on most wall heaters and baseboard heaters. As a rule, wall-mounted line-voltage thermostats are most sensitive to ambient-surrounding temperatures than built-in types.

3. Single-Pole Low-Voltage Thermostat (Figure 7.18).
The line or service voltage is applied across the primary winding of a transformer within the relay and less than 30 volts is impressed across its secondary winding. The thermostat is connected to this low-voltage circuit.

When the thermostat contacts close, current flows through a resistance wire that activates a bimetallic element, which, in turn, operates a switch, and the switch closes. The switch is connected into one side of the service line. When its contacts close, the heater load is energized.

Opening the thermostat contacts disrupts current through the resistance wire. The bimetallic element cools and operates the switch so that the contacts open to disrupt current to the heating load.

4. Double-Pole Low-Voltage Thermostat (Figure 7.19).
Just as in the single-pole unit, a step-down transformer delivers about 30 volts to the thermostat.

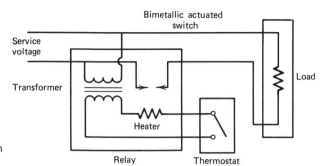

Figure 7.18 Low-voltage thermostat with relay.

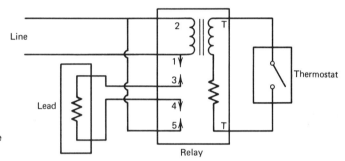

Figure 7.19 Low-voltage double-pole thermostat with relay.

When the thermostat contacts "make," the voltage energizes a solenoid coil that pulls the mercury switch over and closes their contacts. Opening the thermostat contacts de-energizes the coil, and a spring forces the mercury switches to assume their normal open-contacts position.

Low-voltage thermostats are less susceptible to "droop" than line-voltage thermostats. Since the heating load requirements are handled by the relay, the size of the load does not affect the operation of the thermostat. Moreover, the thermostat is less susceptible to contact arcing and welding, since the unit is required to "make" and "break" less voltage.

Code Standards The following section from the NEC applies to thermostats. 424-20 Controllers and Disconnecting Means.

(a) Thermostats and thermostatically controlled switching devices that indicate an off position and interrupt line current shall open all ungrounded conductors when the control device is in this off position.

(b) Thermostats and thermostatically controlled switching devices that do not have an off position shall not be required to open all ungrounded conductors. Remote-control thermostats shall not be required to meet the requirements of (a) and (b) above. These devices shall not be considered as a disconnecting means.

Thermostat Installation The location of the thermostat is an important consideration in any type of heating.

There are five basic rules:

1. Mount it on an inside wall approximately five feet from floor level.
2. Keep the thermostat away from all direct heat and direct light sources. That means don't expose it to the sun or a lamp, and don't put it near a fireplace or a TV set. Heat from such sources will activate the sensing element and cut off the heater when the ambient temperature in the room is below the control point.
3. Don't install the thermostat near ductwork or piping in the wall for the same reason. The unit won't accurately record ambient temperatures in the room.
4. When mounting the thermostat on the wall, use a spirit level or a plumb bob to insure a vertical position. The calibration does not hold true unless the unit is perfectly level.
5. Finally, when location is desired, it is a good practice to drive a 16 penny spike into the nearest 2 X 4 wall stud. Wrap the thermostat wire around the spike several times. This will mark your location, as the sheetrock installers will have to cut a hole in the sheetrock at this point. Not only do you have your location marked out, you also have sufficient backing for mounting the thermostat.

7-7 Determining Heating Requirements To determine the heating requirements and design of the heating system, it is essential to calculate the rate of heat loss per hour so that a comfortable environment will be maintained for the activity inside the house.

The temperature to be maintained can vary to conform to the needs and activities of the occupants. For the healthy active child, 60° might be comfortable, while 80° might be more suitable for the older person. Naturally, the definition of comfort will vary from person to person and from one type of activity to another. We therefore have to make our analysis based on average types of usage.

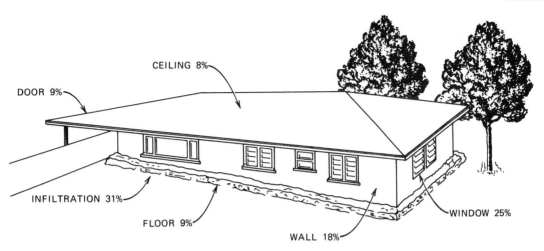

DOOR 9%

CEILING 8%

INFILTRATION 31%

FLOOR 9%

WALL 18%

WINDOW 25%

Figure 7.20 A typical house with adequate insulation, showing heat loss by conduction through the structure. (Courtesy of Martin Industries)

Where Do We Lose Heat?

Look at a typical house, shown in Figure 7.20, and see where the heat escapes. (This is an example of contemporary houses with adequate insulation.)

What Determines Heat Loss?

The total heat loss from a house can be divided into two major parts:

1. The heat loss by conduction through the structure.
2. Heat loss through air infiltration. All homes have some air exchange — cold air entering the house and warm air leaving. This is called infiltration.

    This cold air must be heated and constitutes a significant load on the heating system.

    The type of structure and its insulating values can greatly affect the amount of heat loss. You can see from the house that glass areas lose more heat than insulated areas.

Insulation

Insulation is any material that reduces the rate of heat transfer from one area to another. All building materials have some insulation value, but the term insulation is generally applied to a group of products designed mainly to provide this one service.

Insulation performs several functions:

1. It helps conserve heat during cold weather and therefore reduces the energy required to heat the home.
2. It helps reduce the rate of heat gain in hot weather, thereby enabling a house to remain cooler inside in summer than it otherwise would.
3. It maintains higher inside surface temperatures. A person loses heat by radiation to warmer surfaces and will be more comfortable. Warmer inside surfaces also minimize condensation or sweating. As an example, double windows collect less condensate than single windows.

Insulation Values One way to compare insulating materials is by thermal resistance (R) or ability to resist the passage of heat. This resistance, determined by standard testing procedures, may be expressed per inch of thickness as manufactured. The higher the R value, the better the insulation properties.

The difference in "R" value can greatly affect the heat loss. One example is in wall and ceiling insulation. A 2 X 4 has an "R" value of 1.25 per inch thickness, where fiberglass has an "R" value of 4.00 per inch thickness. Therefore, the part of a structure that is framing loses more heat.

Infiltration Some air infiltration is necessary in any home. Fuel-burning furnaces require air for combustion and to remove the products of combustion from the house. All homes require some air exchange to remove water vapor that is produced within the home. If the house is extremely tight and there is not enough air exchange, severe condensation may occur on windows and other cold surfaces during cold weather. Infiltration is commonly expressed in air changes per hour. Loosely constructed houses may have an infiltration rate of 2 air changes per hour, whereas tight houses may have an infiltration rate of ½ air change per hour or less. The average home will have an infiltration rate of about 1 air change per hour. Infiltration can be reduced by weatherstripping doors and windows, caulking sills and other openings, keeping fireplace dampers closed, and reducing the operating time of exhaust fans.

Design Temperature The extremes in temperature from inside to outside, called the design temperature, will also determine the amount of heat required to make a space comfortable.

The design temperature varies from area to area. The table below will give a guideline in selecting the proper value.

Table 7.1.
Winter Outdoor Design Temperatures

Degrees Fahrenheit	Degrees Fahrenheit
40 Miami Beach, Florida	−20 North Platte, Nebraska
30 Southern Texas	−20 Northern Maine
20 San Diego, California	−20 Helena, Montana
10 Arkansas	−20 Minneapolis, Minnesota
0 New York City	−25 LaCrosse, Wisconsin
−10 Kansas	−30 Bismarck, North Dakota
−10 Chicago, Illinois	−35 Miles City, Montana
−15 Des Moines, Iowa	−40 Regina, Saskatchewan

Terminology As in any analysis, there are certain terms that are used and should be understood:

BTU: British Thermal Unit (the amount of heat required to raise 1 pound of water 1° F in 1 hour).
Watt: 1 kWh = 3.413 BTU.
"U": The units of BTU per hour loss per square foot per degree F temperature differential.
"R" Value: The amount of heat a material will hold back. The resistance to heat transfer expressed in total "R" or "R" per 1″ thickness:
R = 1 / U

7-8 Calculating Heat Loss It has been shown that the amount of heat loss is determined by:
Design Temperature
"R" Value of the surface
Area of the surface
Putting these into a formula, we have the following:

$$\text{Heat Loss (Watts)} = \frac{A \times DT}{\text{"R"} \times 3.413}$$

A— Area of exposed surface

DT— Design temperature (indoor to outdoor)

"R"— "R" Value for surface (resistance to heat flow)

3.412— Conversion from BTU to watts per hour.

To simplify it, we have worked out multipliers in watts/sq. ft. This multiplier includes the design temperature, "R" value, and 3.412 conversion factor. All you have to do is multiply by the area of a surface. The multipliers in Table 7.2 are for typical construction. They naturally do not include every possible situation, and multipliers can be developed for any situation by using the "Heat Loss Formula" above.

Table 7.2

Heat Loss Multiplier (In Watts per Square Foot)

	"R" Value	Design Temp. Differential			
	(Average)	70	80	90	100
Walls (Frame)					
3½" Fiberglass (10% framing)	14.6	1.4	1.6	1.8	2.0
3½" Fiberglass (10% framing) (w/1" Rigid Insul. overall)	18.6	1.1	1.3	1.4	1.6
(Block)					
8" No Insulation	1.04	19.7	22.5	25.3	28.2
12" No Insulation	1.56	13.1	15.0	16.9	18.8
w/2" Insulation	9.56	2.1	2.5	2.8	3.1
Ceiling					
4" Fiberglass (10% framing)	14.3	1.4	1.6	1.8	2.0
6" Fiberglass (10% framing)	21.7	0.94	1.1	1.2	1.3
8" Fiberglass (2" over framing)	29.1	0.7	0.8	0.9	1.0
10" Fiberglass (4" over framing)	36.5	0.56	0.64	0.72	0.8
Windows					
Single Glass	0.89	23.0	26.3	29.6	32.9
Single Glass w/Storm	1.79	11.4	13.1	14.7	16.4
Double Pane, ½" Air Space	1.73	11.8	13.5	15.2	16.9
Triple Pane, ¼" Air Space	2.12	9.7	11.1	12.4	13.8
Doors					
2" Solid, No Glass	2.5	8.2	9.4	10.5	11.7
2" Solid, w/Storm	4.6	4.5	5.1	5.7	6.4
Floors (Over crawl space) (40° min. temp.)					
No Insulation	1.25	16.4	18.7	21.1	23.4
2" Fiberglass	9.25	2.2	2.5	2.8	3.2
4" Fiberglass	17.25	1.2	1.4	1.5	1.7

Table 7.2 (Continued)
Heat Loss Multiplier (In Watts per Square Foot)

	"R" Value (Average)	70	80	90	100
		Design Temp. Differential			
(Concrete slab)[1]					
No Insulation	1.2	17.0	19.5	11.0	24.4
2" Rigid Insul. 4' Down	9.2	2.2	2.5	2.9	3.1
Rim joist area[2]					
No Insulation	3.0	6.8	7.8	8.8	9.8
4" Fiberglass	19.0	1.1	1.2	1.4	1.6
Infiltration[3]					
Tight Construction (½ air chg/hr)		0.35	0.4	0.45	0.5
Loose Construction (1 air chg/hr)		0.7	0.8	0.9	1.0
Fireplace[4]					
Tight Damper		410	470	525	586
Loose Damper		1022	1168	1314	1460
No Damper		3070	3512	3951	4390

[1] *Heat loss on slabs is based on perimeter of slab exposed to the outside; therefore it is in lineal feet.*
[2] *Rim Joint Area is the framing space between floors and concrete block walls—if uninsulated, can be an area of high heat loss.*
[3] *See discussion of Infiltration on page 116.*
[4] *Fireplace losses are expressed in watts for each Design Temperature. Add these values to your normal calculation for each room with a fireplace.*

Summary
1. With electric space heating, temperature requirements are met quickly, quietly, and without chilly drafts.
2. A study dwindling availability of fossil fuels has caused a growth in the use of electrical energy.
3. Electric space heating is foolproof and maintenance-free.
4. Thermostats are installed on inside walls away from ductwork and piping.
5. Inside temperatures can vary to conform to the needs and activities of the occupants.
6. The type of structure and its insulating value will effect the amount of heat loss.
7. Insulation reduces heat transfer from one area to another.
8. Electric heat panels are ideal for an area where a concentrated heat is needed.
9. Infiltration is expressed in air changes per hour.

Sample Calculation

Design temp._____° diff.

Room_____		Multiplier (10)	Loss watts/hr (11)	Notes—htg. equip.

Size_____×_____×_____

Outside wall gross	_____ sq. ft.	_____	_____	_____
Windows	_____ sq. ft.	_____	_____	_____
Doors	_____ sq. ft.	_____	_____	_____
Outside wall net	_____ sq. ft.	_____	_____	_____
Ceiling	_____ sq. ft.	_____	_____	_____
Floor (wood over air)	_____ sq. ft.	_____	_____	_____
Floor (concrete slab)	_____ lin. ft.	_____	_____	_____
Air change	_____ cu. ft.	_____	_____	_____

TOTAL HEAT LOSS FOR THIS ROOM _____(12)_____

1. **Design temperature**—Select from Chart. In Minneapolis 90°F is normally used.
2. **Room and size**—Identify the room or area.
3. **Outside wall gross**—This is total wall area exposed to the outside.
4. **Windows**—Total window area in the room. Remember, sliding glass doors are the same as windows in calculation.
5. **Doors**—Include any glass area in doors in (4) above.
6. **Outside wall net**—Subtract door and window areas from outside wall gross. This gives you the insulated wall area.
7. **Ceiling**—If there are any skylights, they must be considered separately.
8. **Floor**—Over heated basement, no loss. Unheated or crawl space depends upon insulation. If it is a concrete slab, it depends on lineal footing around the perimeter, rather than square footage of area.
9. **Air change**—Depends upon how tight the structure is and if it has a fireplace or not. Remember how tight the damper is on the fireplace could also affect this. *Note:* This will be based on cubic footage or volume, not area.
10. **Multiplier**—After you have put in the areas or volume, insert the multiplier from Page 118 or from your own calculation.
11. **Loss in watts/hr**—Multiply them out and you can determine how much watt loss you have from each part of the structure.
12. **Total heat loss for this room**—Add these up and it gives you a minimum size of heating unit in watts that is required to make up for the heat loss in this room.

Problems 7-1 List at least six advantages that electric space heating has over other heating systems.

7-2 List seven different types of heating units available for installation in a residence.

7-3 Explain what is meant by convection heating.

7-4 Draw a wiring diagram of a baseboard heater being controlled by a double-pole line voltage thermostat.

7-5 Explain what is meant by the R designation of an insulation.

7-6 Explain heat loss through air infiltration.

Chapter 8 Solar Energy

Instructional Objectives
1. To understand the need for harnessing the sun's power.
2. To make you aware of our environment.
3. To learn why we need other sources of energy.
4. To provide the basic components in the solar system.
5. To learn some advantages of the storage system.
6. To understand the operation of heat transfer.
7. To become familiar with the types of collectors.
8. To make you aware of the need for auxiliary energy.

Self-Evaluation Questions

Test your prior knowledge of the information in this chapter by answering the following questions. As you read the chapter, watch for the answers. When you have completed the chapter, return to this section and answer the questions again.

1. Why do we need to find ways to harness the sun's power?
2. What types of fuel are in short supply?
3. Will solar energy be used for more than space heating?
4. How is energy collected?
5. Why is a good storage system essential?
6. What will happen to solar energy on a cloudy day?
7. Is solar heat effective in a cold climate?
8. What is the purpose of adding auxiliary energy?
9. Where will the auxiliary system be located?
10. What is the purpose of the control component?

8-1 Energy From The Sun

This section is about solar energy and how it is used for space heating, cooling, and domestic hot water heating. It has been prepared as an introduction for the newcomer, apprentice, and owner or builder interested in reducing heating costs by harnessing the sun's power.

The United States currently faces two interrelated problems. One, we are endangering our environment by polluting the atmosphere, ground, and water with the by-products of our techno-

123

logical society. Two, we are quickly running out of fuels that have enabled us to achieve a high standard of technological development. This perplexing dilemma has resulted in a nationwide investigation of energy choices.

Among the many energy alternatives being considered is solar energy. Harnessing the sun's power is considered an attractive alternative because it is a renewable resource that does not pollute. In contrast to conventional fuels, its use eliminates the need for refining, transportation, and conveying fuel and power over long distances. The use of solar energy for heating and cooling promises a more rapid payoff than other energy alternatives, because the basic technology already exists and needs only minor refinement. Considerable research, development, and demonstration activities have been initiated in the public and private sector to facilitate the widespread utilization of solar energy.

8-2
Solar System Components

Several characteristic properties apply to all solar heating, cooling, and domestic hot water systems, whether they are simple or relatively complex. Any solar system consists of three generic components: collector, storage, and distribution, and may include three additional components: transportation, auxiliary energy system, and controls (Figure 8.1). These components may vary widely in design and function. They may, in fact, be one and the same element (a masonry wall can be seen as a collector, although a relatively inefficient one, that stores and then radiates or "distributes" heat directly to the building interior). They may also be arranged in numerous combinations dependent on function, component compatibility, climatic conditions, required performance, and architectural requirements.

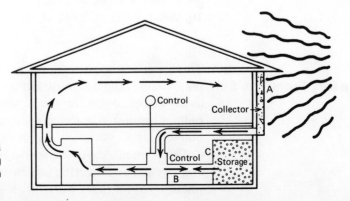

Figure 8.1 Solar system components; *A* Taking heat from collector; *B* Storing heat from collector; *C* Heating from storage.

The Collector The collector converts incident solar radiation to usable thermal energy by absorption on a suitable surface. The thermal energy captured is transformed to a heat transfer medium, usually gas or liquid within the collector.

There are numerous concepts for the collector of solar radiation. These concepts range from the most simple—a window—to those that are quite complex and require advanced technology for their development.

Collectors are generally classified as focusing or nonfocusing, depending upon whether the sun's energy is concentrated prior to being absorbed or collected at the density received at the earth's surface.

Of the many solar heat collection concepts presently being developed, the relatively simple flat-plate collector has found the widest application. Its low fabrication, installation, and maintenance cost as compared to higher temperature heat collection shapes has been the primary reason for its widespread use.

Flat-plate collectors utilize direct as well as diffuse solar radiation. Temperatures to 250° F (121° C) can be attained by carefully designed flat-plate collectors.

A flat-plate collector generally consists of an absorbing plate, often metallic, which may be flat, corrugated or grooved; painted black to increase absorption of the sun's heat; insulated on its backside to minimize heat loss from the plate; and covered with a transparent cover sheet to trap heat within the absorber. The captured solar heat is removed from the absorber by means of forced air circulating underneath the collectors. A fluid such as water or air passes through the collector, Figure 8.2, picks up the sun's heat from the hot absorber surface and transports that heat away from the collector. The heated air or liquid can be used directly or it can give up its heat to a storage container (Figure 8.3).

Storage The storage component of a solar system is a reservoir capable of storing thermal energy. Storage is required since the largest portion of total heat usage will occur at night or on consecutive sunless days when collection is not occurring. Storage acquires heat when the energy delivered by the sun and captured by the collector exceeds that demand at the point of use.

Air systems generally have a large insulated storage tank consisting of washed river rock about the size of a golfball. The rock storage is heated as the air from the collector is forced through the rock container by a blower. A decrease in rock size increases the air flow

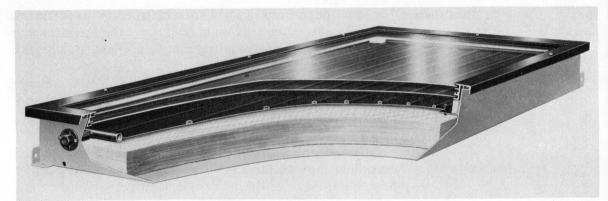

Figure 8.2 Flat-plate solar collector; with a high-absorption, black chrome coated absorber plate, an acid-etched, tempered low iron glass cover for increased light transmission and rugged, fully insulated construction. (Courtesy of Lennox Industries, Inc.)

resistance through the storage and may affect blower and duct size and distribution efficiency. Rock storage does not have to be in close proximity to the collector. However, as the distance increases, the heat transfer losses between the heated air and the rocks also increase, and larger ducts and more electrical power are generally required for moving air between the collector, storage, and heated spaces.

Domestic hot water piping is run through the storage tank prior to passing through a conventional water heater. Storage heat is transferred to the hot water piping, thereby either eliminating the need for additional heating or substantially reducing the energy required to raise the water to the needed distribution temperature.

Distribution The distribution component receives energy from the collector or storage component and disperses it at points of consumption— spaces within the dwelling. For example, comfort heat is usually distributed in the form of warm air or warm water by ducts or pipes within a building. Distribution of energy will depend upon the temperature available from storage space heating if baseboard convectors are increased in size or if used in conjunction with a heat pump or auxiliary heating system.

Because solar-produced temperatures in storage are normally in the low range (90 to 180° F), distribution ducts and radiating sur-

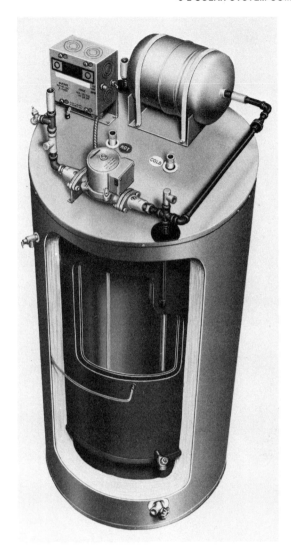

Figure 8.3 A solar heat storage tank, a heat exchanger, and control center all in one. A heat transfer fluid is circulated through the collectors where it absorbs heat. This fluid is then piped to the built-in heat exchanger in the storage tank. The cooled transfer fluid is then pumped back to the collectors to continue the cycle. As hot water is drawn from the conventional water heater, hot water from the solar module storage tank replaces it. (Courtesy of Lennox Industries, Inc.)

faces are normally larger than those used in conventional heating systems. Therefore, careful consideration is required in the design of heat distribution systems throughout the dwelling.

Domestic water heating is also a part of the distribution component. Its distribution system generally consists of a heat exchanger, back-up heater, piping, and controls.

Transport Most solar systems have an energy transport component that provides the means of moving a fluid carrying thermal energy to and from the collector and storage. The transport component also regulates the flow through the collector and storage. In liquid or air systems this component consists of pumps, valves, and pipes, or blowers, dampers, and ducts.

Auxiliary Energy The auxiliary component provides a supply of energy for use during periods when the solar system is inoperable or during periods of extremely severe weather or extended cloudy weather when solar-produced temperatures from the collector and storage are not sufficient to satisfy the dwelling's heating or cooling load. Currently, the experimental nature of solar heating, cooling, and domestic hot water systems and the possibility of extended sunless days generally require that the auxiliary energy components be capable of providing for the total energy demand of the house if the solar system is inoperative.

Almost any type of auxiliary energy system may be used in conjunction with a solar system. The auxiliary system may be completely separate or fully integrated with the solar heating/cooling system. However, in most cases, it makes economic sense to integrate the back-up system with the solar system. This means running the distribution component from heat storage to the occupied space through the auxiliary system where an energy boost may be supplied when storage temperatures are low. Heat from storage may also be used in conjunction with heat pumps. The heat pump, a device that transfers heat from one temperature level to another by means of an electrically driven compressor, utilizes the solar heat available from the storage to supply necessary heat to the occupied space. The advantage of the heat pumps/solar system integration is the reduction of electrical energy required by the heat pump because of heat supplied by solar storage. (Heat pumps are discussed in Chapter 9.)

Supplemental heat may also be supplied by installing electrical resistance duct heaters into the duct work of the system. (Refer back to Chapter 7 and types of heating units.)

Control The control component performs the sensing, evaluation, and response functions required to operate the system in the desired mode. For example, the temperature in the house is sensed by a thermostat and relayed to the distribution component (pump or blower) when heat is required. The controls generally distribute information, including fail-safe instructions, throughout the system by means of electrical signals. However, the control function can be performed by automatic pneumatic controls or by the dwelling occupants who initiate manual adjustments to alter the system (Figure 8.4).

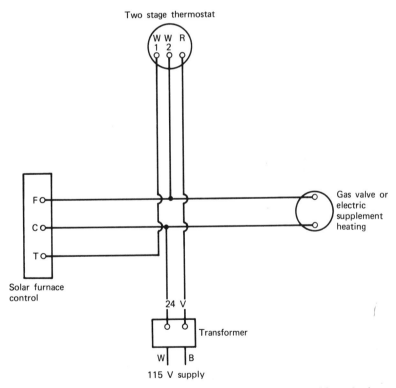

Figure 8.4 Wiring diagram showing two-stage thermostat control for solar furnace tied in with gas valve on furnace or electric supplement heating.

Summary **1.** A solar system consists of three generic components: collector, storage, and distribution.

2. The collector converts solar radiation to usable thermal energy by absorption.

3. The storage component of a solar system is a reservoir capable of storing thermal energy.

4. An auxiliary system can be powered by conventional fuels or electricity.

5. The control component performs the sensing, evaluation, and response functions required to operate the system.

Problems **8-1** List the six components required for a solar heating system.

8-2 Explain how conventional fuels or electricity could be added to supplement heat on a cloudy day or cold night.

8-3 Draw a sketch of a solar system showing a collector storage and distribution of air to a single-story house.

8-4 Draw a one-line drawing of the control system, showing a two-stage thermostat hooked into a solar furnace.

8-5 Explain why a good storage system is essential.

Chapter 9 The Heat Pump

Instructional Objectives
1. To become familiar with the basic operation of the heat pump.
2. To make you aware of the alternatives.
3. To become familiar with the outdoor compressor.
4. To understand why the heat pump is sized for the cooling load of the house.
5. To learn the advantages of the heat pump.
6. To develop an understanding of efficiency loss.
7. To understand unit maintenance of the system.

Self-Evaluation Questions

Test your prior knowledge of the information in this chapter by answering the following questions. Watch for the answers as you read the chapter. Your final evaluation of whether you understand the material is measured by your ability to answer these questions.

1. What is a heat pump?
2. How does the heat pump transfer heat/cooling to the rooms?
3. How is the refrigerant reversed?
4. What is needed during extreme cold weather?
5. How is heat/cooling controlled?

9-1 The Heat Pump

In these days of lower thermostat settings and revamped heating systems, the heat pump is gaining attention as a possible energy-saver, full-efficient alternative to the more traditional home heating systems.

Actually, the heat pump may sound new, but it's not. It was used experimentally in the 1930's and was adopted by the United States Army during World War II for use in some government buildings in the southern United States.

It wasn't until quite recently that the technology of heat pumps had advanced to the stage that they could be considered as viable heating alternatives in more severe northern climates.

Today, manufacturers say, heat pumps are in use as far north as southern Canada.

HOW A HEAT PUMP WORKS

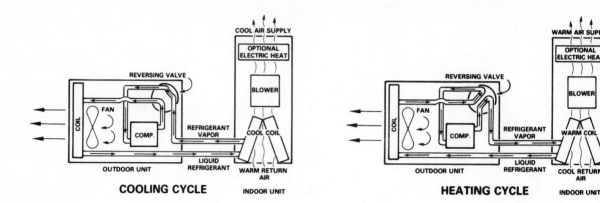

Figure 9.1 Operating principles of the heat pump. (Courtesy of Lennox Industries, Inc.)

Cools in Summer

During cooling, a heat pump removes heat from indoor air by circulating liquid refrigerant from the outdoor coil to the indoor coil. Here, warm indoor air is forced over the coil surface by a blower. Liquid refrigerant in the coil changes to cold vapor and absorbs the heat. Air flowing over the finned coil is cooled, excess moisture is condensed. This cool, dry air is then circulated gently and quietly throughout your home.

The refrigerant vapor, having absorbed heat, is returned to the outdoor unit where the compressor and outdoor coil convert it back into a liquid and discharge the heat. The liquid refrigerant then travels back indoors to continue the cooling process as long as there is a need.

Heats in Winter

During heating, a reversing valve changes refrigerant flow so the heat pump can use solar energy present in outdoor air to heat your home. Refrigerant in the outdoor coil absorbs heat from air passing over it. Even at 0°F, air still contains over 82% of the heat that was available at 100°F. (That may seem strange but it's true.) The compressor pumps the refrigerant, now in a hot vapor form, to the indoor coil. The blower circulates indoor air over the hot coil, warming it for distribution throughout your home. As the hot vapor cools, it condenses and the resulting liquid refrigerant returns to the outdoor coil. There it once again absorbs heat and repeats the cycle as long as heating is needed.

**9-2
The Heat Pump Defined** What is a heat pump? To oversimplify, the heat pump is an electrically operated system that can both heat and cool the conditioned area (Figure 9.1).

During the heating cycle, heat is "pumped" from outside air through coils containing a pressurized liquid (there is heat in cool outside air). The heated liquid is then vaporized and, as the inside air circulates around the coils, the heat is transferred to the air and pumped into the house heating-ducts.

In the summer months, this operation is reversed. The heat pump reverses the flow of heat by using a system of valves in the refrigerant that turn the refrigerant around and make it flow in the opposite direction. The pressurized liquid is pumped into the indoor coils where warm air is blown over them. The coils absorb the heat from the air and pump it to the outside. The cooled indoor air is then circulated through the house.

Heat pumps distribute the heat or cooling to the rooms through a central duct system in the same manner as other central forced-air systems (Figure 9.2).

It should be noted, however, that the heat pump is usually sized for the cooling load of the house. If heat pumps were sized to handle total heating demands in homes during the coldest winter weather periods, they would be considerably oversized with relation to summer cooling needs.

**9-3
Advantages of the
Heat Pump
Economy**

When heating, the aim of the reverse cycle system is to get as much heat as possible from every unit of energy used. When we burn a gallon of fuel oil, we want to get as many BTU's as possible from the gallon. Any heat that goes up the chimney is loss.

The fuel used for heating with heat pumps is electricity. Whenever you convert electricity to heat with ordinary electric resistance heaters, you do this heating at a 100 percent efficiency. For every one cent of electricty you buy, you get one cent of electricity back in the form of usable heat.

The great advantage of the heat pump is that for every cent of electricity that you buy, you will get back considerably more than one cent's worth of electric heat. With a well-designed air-to-air heat pump, you will usually get back three cents' worth of electric heat for every one cent of electricity you buy to run the heat pump in 40° winter weather. The two cents of electric heat that you get for free, in this example, have been pulled out of the outdoor air by the heat pump.

The ratio between the heat output and the heat (or electric) input to the heat pump is known as the COP, which stands for "coefficient of performance."

One kilowatt hour of electricity contains 3,413 BTU. If the COP of a heat pump is 3.00, then, for every kilowatt hour that you spend to run it, you will get three times this, or 10,239 BTU's of heat delivered from the machine.

As the weather gets colder, the COP of the heat pump goes down so that you get less and less heat out of the air "for free."

(a)

(b)

Figure 9.2 The heat pump. (a) Outdoor unit of the H.P. 10 series heat pump, suitable for rooftops or grade level slab installations. Note that the compressor is housed in the outdoor unit as in conventional split cooling sets. In the heat pump, however, the function of the outdoor and indoor coils reverses seasonally by means of a reversing valve. (b) Typical indoor blower-coil-filter unit, extremely versatile, can be installed in a closet, attic space, suspended from a ceiling or in a crawl space. (Courtesy of Lennox Industries, Inc.)

Service Contract An obvious advantage to these reverse-cycle air-conditioning systems is that, since the same equipment serves both heating and cooling requirements, maintenance responsibilities for both functions can be handled by a single service contract.

Health Another advantage is that the filtering and evaporative facilities free the interior air of excess moisture, dust, and pollen, permitting persons who normally suffer from hay fever or airborne allergies to obtain relief in the tempered atmosphere.

9-4 Alternatives The heat pump is a possible alternative to natural gas, oil, and electrical resistance heating systems.

As stated in other chapters, the uncertain future of the availability of natural gas and heating oil has focused attention on electrically operated systems; and of the two basic systems mass-marketed today, electrical resistance heating and the heat pump, the heat pump shows more promise in terms of energy conservation. (Solar energy is not mentioned here due to the lack of mass production of the collector and because of comparable energy saving statistics.)

9-5 Efficiency Loss The heat pump is usually sized for the cooling load of the house. As a reult, the heat pump cannot fully satisfy a home-owner's needs in severe winters. When the outside temperature drops to below freezing, the heat pump starts to lose its heating efficiency. Heat pumps, at least in northern climates, require a supplemental heat source such as the standard basement gas, oil forced-air furnace, or electrical resistance heating. As mentioned in Chapter 7, supplementary electric resistance duct heaters could easily be installed in the duct work to make up for heat pump limitations during such extreme cold snaps.

9-6 Control Heating and cooling functions are controlled by a two-dial, 24-volt summer/winter thermostat in the living room, one dial for heating and one for cooling (Figure 9.3). A four-position switch on the thermostat sets its function to heat control, cooling control, fan operation, or off. Individual room temperatures may be adjusted by means of register dampers in the duct system.

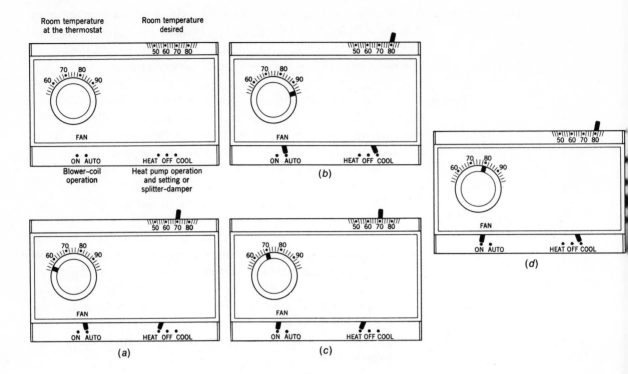

Figure 9.3 Thermostat for the control of heat pump and blower coil unit. (a)Turn on for heating. Room temperature 55° F. Desired temperature set at 70° F. Heat pump set for *heat*. Blower set for *automatic*. When the heat pump starts, blower turns on automatically. When room temperature reaches 70° F, both turn off. (b) Turn on for cooling. Room temperature 95° F, desired temperature 80° F. Blower set *automatic* and heat pump at *cool*. Blower follows heat pump automatically. When room temperature get down to 80° F, both turn off. (c) Continuous circulation—heating (d) Continuous circulation—cooling. (Courtesy of Lennox Industries, Inc.)

9-7 Maintenance Because the pad-mounted compressor is located outside the house, there is ample room for maintenance.

Maintenance requirements are mainly a matter of periodically replacing filters, lubricating bearings, checking loose fan belts, keeping electrical connections tight, occasionally replacing faulty relays, and keeping contacts clean and in positive-action.

Summary 1. A heat pump is year-round comfort control from only one system.
2. Electricity is the only power/fuel source needed for the operation of the heat pump.

3. With a heat pump, there is only one system to service.
4. A heat pump may be tied in with solar energy.
5. Heating and cooling are controlled by a two-dial, summer/winter thermostat.
6. Supplemental resistance heaters may be installed for extreme cold winter days.

Problems 9-1 Explain briefly how a heat pump works: (a) in winter and (b) in summer.

9-2 List three advantages of using the heat pump.

9-3 Why is the heat pump sized for cooling load rather than for heating?

9-4 Draw a wiring diagram of a heat pump control thermostat.

9-5 List six maintenance requirements for the heat pump.

Chapter 10

**Forced
Air
Heating**

1. To understand the operation of forced air heating.
2. To learn how warm air is circulated.
3. To learn why a perimeter type duct installation is installed for forced warm air furnaces.
4. To become familiar with the types of fuel used in today's residential heating.
5. To learn why some home owners heat with a different type of fuel.
6. To understand the need for automatic controls.
7. To review the operation of the thermostat.
8. To understand the operation of the fan control.
9. To learn some advantages of installing an electronic air cleaner.
10. To become more familiar with the operation of the air cleaner.

**Self-Evaluation
Questions** Test your prior knowledge of the information in this chapter by answering the following questions. As you read the chapter, watch for the answers. When you have completed the chapter, return to this section and answer the questions again.

1. What is the purpose of the mechanical blower?
2. Why is it important to check on the fuel before installing a forced air heater?
3. Why is natural gas a popular fuel?
4. What other fuels are used besides natural gas?
5. What does the control system do?
6. How is a diaphragm valve actuated?
7. What is a stock-relay switch?
8. In an air burner installation, how is the oil ignited?
9. Where is the fan control located?
10. Give one good reason for using a room thermostat.
11. What does an electronic air filter do?

10-1
Forced Warm Air
Furnaces

Today, most furnaces being installed circulate warm air by mechanical means and are often referred to as forced air furnaces.

Forced warm air furnaces include a fan or blower as part of the unit. The purpose of the blower is to circulate the heated air to all of the rooms. Since a mechanical blower circulates the warm air, the heat ducts may be installed either in a vertical or horizontal direction.

Most forced air furnaces contain an air filter to remove impurities in the air. An electric motor, along with proper pulley and belt, drives the forced air blower. The blower forces the air across the heat exchanger up into the plenum and out through the ducts and pipes to the various rooms.

Today, the perimeter type of duct installation is used with the forced warm air furnace. This system distributes warm air to the perimeter of the house (Figure 10.1).

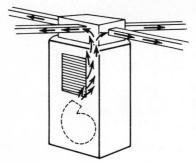

Figure 10.1
Furnace cabinet and plenum.

10-2
Fuel Considerations

The selection of a fuel depends on many factors. However, in most localities, we need only to choose between two, gas (which includes natural, mixed manufactured and liquified petroleum gases) and fuel oil. Some of the considerations center on economics, such as the cost of the fuel or initial cost of the heating system adapted for that particular fuel. Other considerations might be the availability of the supply, the convenience of handling the fuel, and local ordinances and zoning regulations that may prohibit the use of certain fuels.

Natural Gas

The most popular fuel being used today is natural gas, since it is used in far more new installations than any other fuel (Figure 10.2).

A vast network of gas pipelines reaches a large part of the country. These pipes bring natural gas from the fields and often store it in underground locations conveniently located with relation to the gas markets. In many parts of the country, natural gas is an economical fuel to burn. Furnaces using gas are generally lower in initial cost than equipment using other fuels. The efficiency of gas-burning equipment is relatively high, and there is no requirement for fuel-storage facilities on the premises where natural gas is available.

In some parts of the country, natural gas is not available and manufactured gas is used. Manufactured gas may be made from

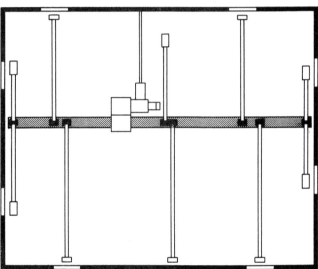

Figure 10.2 Forced-air gas furnace. GS 11 Series. (Lennox Industries, Inc.)

coal or converted from fuel oil, and it is usually more expensive than natural gas. Nevertheless, it is used extensively where natural gas lines do not exist. In some places, manufactured and natural gas are mixed together to form a highly satisfactory fuel.

In many rural areas where natural gas lines do not exist, liquified petroleum gas (L.P.) or bottle gas is used. This may be either propane or butane gas. In the colder climates, propane is used because butane will not vaporize at temperatures below freezing. With the L.P. gases, it is necessary to store the fuel in tanks located

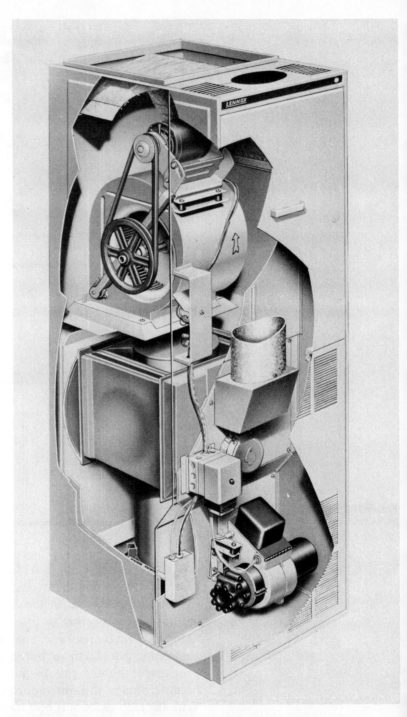

Figure 10.3 Fuel-oil furnace. 0 11 R Series. (Lennox Industries, Inc.)

on the premises but out of doors. These are pressure tanks that hold a combination of liquid and vapor.

Gas furnaces that use liquified petroleum gases are especially designed for L.P. fuel. However, these furnaces may usually be converted rather readily to use natural gas if it becomes available at a later date. Liquified petroleum gases are usually more expensive than natural gas. As a result, many consumers would prefer to have natural gas available as a fuel.

Fuel Oil Probably the second most popular residential heating fuel is oil. However, oil equipment is generally more costly than gas, and it also requires fuel storage equipment, which adds to the installation expense (Figure 10.3).

Oil is, however, a very acceptable residential heating fuel and it is used in many parts of the country, especially where natural gas is not available or where the rate structure is such that gas is more expensive.

Many new homes may have to burn oil because natural gas is not available to them.

10-3 Automatic Controls

Like other controls, the control system performs those functions: safety, comfort, and convenience, in that order of importance.

The main factor to bear in mind when installing automatic controls for gas burners is that a dangerous quantity of gas must not be permitted to accumulate in the combustion chamber, there must be a safety device to stop the flow of gas in case of a pilot failure. There is a variety of safety devices for dealing with this hazard.

In general, there are two means of feeding gas to the burner when the room thermostat calls for heat: one, by means of a diaphragm valve, two, by means of a solenoid-operated valve. The diaphragm valve is pressure-actuated. A small solenoid coil electrically opens a small valve that permits gas to escape from the top of the diaphragm. When this occurs, normal pressure underneath raises the diaphragm, and gas flows to the burner. When a solenoid valve is used, completion of the operating circuit energizes the solenoid coil and pulls up a plunger, which in turn opens the valve (Figure 10.4).

Figure 10.5 shows a layout of the relative positions of the equipment needed for a gas-burner installation. Note that the electric

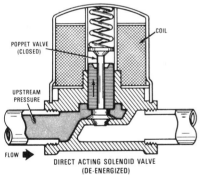

Figure 10.4 Solenoid type main gas valve is opened and closed by thermostat switching on and off magnetic coil that lifts the plunger. Spring closes valve.

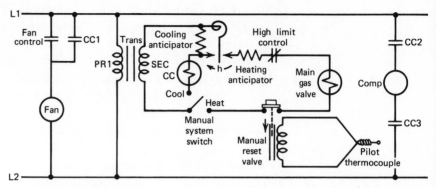

Figure 10.5 Typical control schematic for a gas-fired, warm air heating system with summer cooling added.

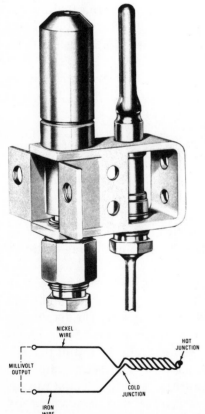

Figure 10.6 The thermocouple converts heat into electricity in sufficient quantities to power safety devices, relays, or valves. Note thermocouple mounted next to a pilot light.

safety pilot is wired to a thermocouple. A thermocouple, Figure 10.6, is made of two unlike metals that generate a small amount of electricity when heated. The current then is used to hold an electromagnetic valve open. In case of pilot failure, the metal would cool. This would stop the flow of current and close the valve.

The installation of oil-burner controls is complicated by the addition of a stack-relay switch. This switch, or protector relay, is installed in the flue pipe. It functions to shut off the burner in case the stack does not come up to a certain predetermined temperature within 45 seconds after the motor starts. The relay prevents flooding the basement with fuel oil in case the ignition system fails.

Ignition of the oil is accomplished with the aid of a spark gap that is connected to the high-voltage terminal of a transformer installed on or near the burner.

The spark may be continuous, or it can be made to cease when the oil starts to burn. The continuous spark system is generally advised when down drafts are common. With a continuous spark, it is necessary to replace the electrode more often than with the intermittent spark.

After the spark has been shut off, the oil continues to burn because the temperature of the firebrick lining is high enough to ignite the oil (Figure 10.7).

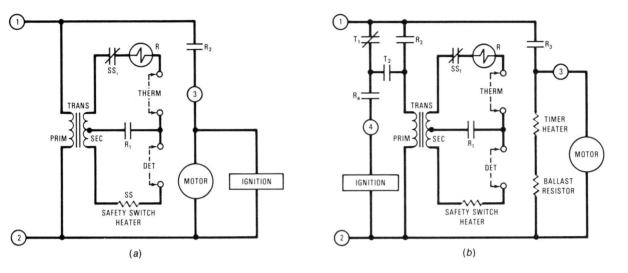

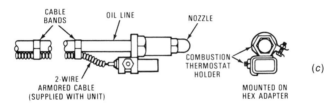

Figure 10.7 Oil-burner controls. (a) Constant ignition wiring hookup. With motor relay contacts closed, ignition is on throughout the call for heat. (b) Intermittent ignition wiring hookup. Timer heater opens contacts and stops ignition after several seconds of successful burner operation. (c) Combustion thermostat (switch) mounted in the burner.

Fan and Limit Controls The function of the fan control is to turn the blower on when the temperature of the air becomes warm enough to be circulated and to turn the blower off when the temperature of the air becomes too low.

The same type fan control can be employed on a heating system regardless of the fuel type used. This control reponds only to the temperature of the air being circulated to the rooms. It is a comfort control, not a safety control. Obviously, if the blower starts to operate before the air has been heated to a sufficiently high temperature, little or no heat will be delivered to the conditioned space. This can result in unpleasant drafts. The same applies if circulation is continued after the air has cooled off excessively following the end of the burning "ON" cycle.

The fan control, a bimetallic element or liquid-filled tube that is exposed to the air circulating over the heat exchanger, opens and closes a line-voltage switch, turning the blower on and off. The fan control is mounted on the furnace cabinet or plenum (Figure 10.8).

Adjustable settings of the fan control are provided by means of two arms. The upper arm is set for the temperature at which the blower should turn on (about 100° F for the high side wall register and 110° F for the low side wall register). The lower arm is set for the temperature at which the blower should stop (about 15° lower than "cut in"). Such fan switches may also have a knob that permits the blower control to be changed from automatic to manual. When the blower control is on manual, the blower runs constantly.

The Limit Control

The limit control is usually combined with the fan switch for warm air furnace applications. In such a combination the limit control responds to the same temperature-sensing element as does the fan switch. When the temperature in the plenum reaches a point above the adjustable setting of the limit control, the limit switch breaks the electrical circuit, thereby causing the burner to stop (Figure 10.8). On a gas burner, the limit switch breaks an electrical circuit, causing the gas valve to close. On an oil burner, the limit switch stops the burner motor. It is common practice to set the limit control "cut out" at about 175° F on a forced warm air furnace.

The limit control never affects the blower operation even when this control is combined with the fan control of a forced air furnace. The blower will continue to operate after the burner has been turned off by the limit control until the plenum temperature has dropped to the cut-out setting of the fan control.

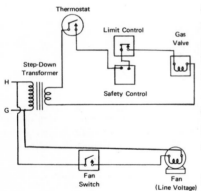

Figure 10.8 Furnace fan and limit control wiring diagram.

The Room Thermostat The room thermostat is used only for regulating room temperature and is in command of the fire if all other controls, such as the limit switch and safety pilot control, are operating properly. For a review of the thermostat, read Chapter 7 — Electric Heat Controls.

10-4 An electronic air cleaner is the newest component that is being
Electronic Air Cleaner installed in cooling and heating comfort-conditioning systems (Figure 10.9).

The electronic air cleaner is capable of removing up to 95% of airborne dirt and irritants, odors, and tobacco smoke, and up to 99% of the airborne pollen present in the air that passs through it.

Basically, the electronic cleaner puts a positive electrical charge in the dust particles that pass through its filter. These particles are

Figure 10.9 Electronic air cleaner. (Lennox Industries, Inc.)

then attracted to negatively charged plates. Some units use a collector cell or filter, which carries small charges of static electricity that attract the dust particles. It is then necessary to use a detergent to wash the electrostatic filter to remove the deposits.

The electronic air cleaner generally has a control that monitors the unit's performance. The control automatically indicates when the filter is dirty and in need of cleaning. Some controls also have timers that permit the serviceperson or homeowner to time the drying cycle after the collector cell or filter has been washed.

Summary
1. In cooler climates propane gas is often used, as butane gas will not readily vaporize at cold temperatures.
2. Natural gas is a common fuel used in residential heating.
3. All electrical controls are switches operated by either a rise or fall in temperature.
4. The limit control breaks on temperature rise, causing the burner to stop.
5. The safety pilot breaks contact on loss of heat on the thermocouple.
6. An electronic air cleaner puts a positive electrical charge in the dust particles that pass through its filter.

Problems
10-1 Name the common fuel types; list advantages and disadvantages of each.
10-2 What factors are considered when selecting a fuel for a residential heating system?
10-3 In the order of importance, list the three functions of the automatic controls.
10-4 Explain how gas is fed to the burner when a room thermostat calls for heat.
10-5 Draw a sketch showing how a safety pilot is wired to a thermocouple.

Part 5 The Service

Chapter 11

**Service
Entrance
Equipment**

**Instructional
Objectives**

1. To develop an ability to identify and define parts of an electrical service equipment.
2. To become familiar with the National Electrical Code as related to service-entrance equipment.
3. To understand the difference between an overhead service and a service lateral.
4. To become familiar with the need for grounding and bonding of the service entrance equipment.
5. To learn why service-entrance conductors are governed by the NEC.
6. To make you aware of the utility company requirements for service laterals.
7. To learn how to ground the meter socket.
8. To understand the need for overcurrent protection against short circuits and faults or overloads.
9. To learn why the grounded neutral conductor is not switched or protected by an overcurrent device.
10. To learn why the power supplier locates the service drop and meter for an owner or builder.

**Self-Evaluation
Questions**

Test your prior knowledge of the information in this chapter by answering the following questions. Watch for the answers as you read the chapter. Your final evaluation of whether you understand the material is measured by your ability to answer these questions. When you have completed the chapter, return to this section and answer the questions again.

1. What is the difference between the service drop and service conductors?
2. Who decides where the service drop shall be located?
3. Why must service equipment be bonded?

4. What is the purpose of a bonding jumper?

5. What is the most desirable electrode for a system ground?

6. What does the term "service" mean?

7. List the main parts of an electrical service installation.

8. Who furnishes the service drop conductors?

9. What type wire can be used for the service-entrance conductors?

10. What is the difference between an overhead service and a service lateral?

11-1
Electrical Service— General

(Figure 11.1)

The electrical service installation is the heart of the electrical system. All electrical energy supplied to power-consuming devices and appliances within the residence must pass through the electrical service entrance equipment where it is metered, protected, and distributed through branch circuits throughout the home (Figure 11.2).

The local power supplier decides where the electrical service will enter the building and where the meter will be located. This is usually the first item to be located before any of the electrical work actually begins. However, we are considering it in this chapter because it is a part of the electrical service equipment.

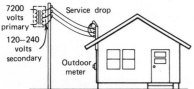

Figure 11.1 Pole mounted, center-tapped secondary, step-down transformer supplying a three-wire, 120/240 volt service to a home.

The service panel should be located near rooms where the largest amount of current will be needed. This is usually in the kitchen.

The term "service" is defined by the National Electrical Code as "the conductors and equipment for delivering energy from the electricity supply system to the wiring system of the premises served."

The main parts of a complete electrical service installation may be listed as: service drop, service conductors, meter, service switch, grounding, and bonding (Figure 11.3). In succeeding paragraphs each of these parts will be defined and their function explained in detail.

11-2
Service Drop

The service drop is actually not part of the wiring system in a residence. It is the connection provided by the power company from their distribution lines to the house. It is discussed in this chapter to familiarize you with the term and its importance.

In Article 100 of the NEC definitions, service drop conductors are defined as "the overhead service conductors from the last pole

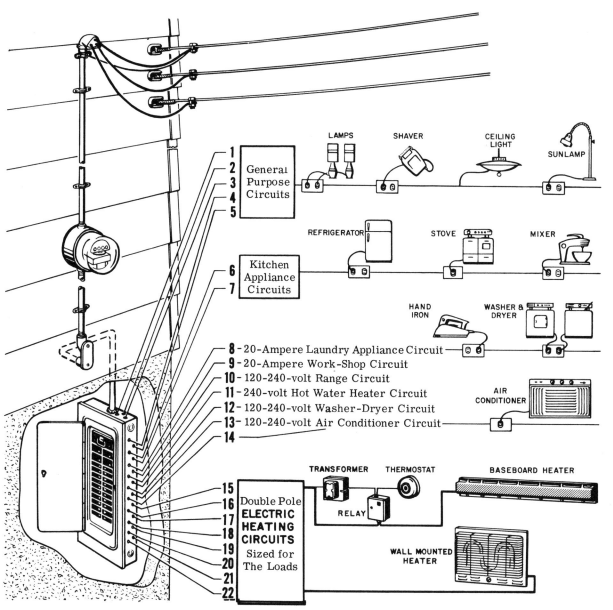

LAMPS   SHAVER   CEILING LIGHT   SUNLAMP

General Purpose Circuits

REFRIGERATOR   STOVE   MIXER

Kitchen Appliance Circuits

HAND IRON   WASHER & DRYER

1
2
3
4
5

6
7

8 - 20-Ampere Laundry Appliance Circuit
9 - 20-Ampere Work-Shop Circuit
10 - 120-240-volt Range Circuit
11 - 240-volt Hot Water Heater Circuit
12 - 120-240-volt Washer-Dryer Circuit
13 - 120-240-volt Air Conditioner Circuit
14

AIR CONDITIONER

TRANSFORMER   THERMOSTAT   BASEBOARD HEATER

15
16
17
18
19
20
21
22

Double Pole ELECTRIC HEATING CIRCUITS Sized for The Loads

RELAY

WALL MOUNTED HEATER

Figure 11.2 Service and Distributions.

or other aerial support to and including the splices if any, connecting the service-entrance conductors at the building or other structure."

Service drop wires are furnished and installed by the power supplier, although the owner or electrician sometimes furnishes the

1. Service drop

2. Entrance head

3. Service entrance conduit

4. Threaded hub

5. Meter base

6. Service entrance panel

7. Main disconnecting means

8. Grounding bushing

9. Equipment bonding jumper

10. Main bonding jumper

11. Grounding electrode conductor

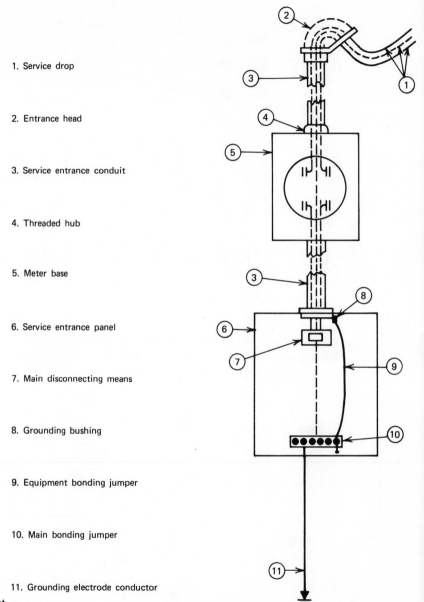

Figure 11.3 Service entrance equipment.

insulators by which the wires are supported on the building or service mast riser. The power supplier also determines the size of the conductors, which are often smaller than the service entrance wires.

The National Electrical Code has several rules governing the installation for the service drop.

1. NEC 230-2   In general, a building or other premises shall be supplied by only one set of service conductors.
2. NEC 230-22 (a)   Individual conductors of multiconductor cable when used as service drop shall be insulated or covered with thermoplastic, rubber or other similar material; except grounded conductor may be bare. (b)   Open conductors shall be insulated or covered and have ampacity as per table 310-17 and 310-19.
3. NEC-230-24 (a.-Ex. 1)   Where voltage between conductors does not exceed 300 and roof has a slope not less than 4 inches, 12 inches, clearance may not be less than 3 feet.
4. NEC 230-24 (a.-Ex. 2)   Service drops 300 volts or less may pass over 4 feet of overhang if minimum of 18 inches clearance is maintained over roof and conductors are terminated in a thru-the-roof raceway or approved support.
5. NEC 230-24 and 230-26   Porcelain insulators for the service drop conductors shall be at least 10 feet above finish grade.
6. NEC 230-28   Service mast shall have adequate strength or be supported by braces or guys.
7. NEC 230-28   All raceway fittings shall be approved for the purpose.

11-3 Service-Entrance Conductors The wires from the point where the service drops end, up to the service equipment, are called service-entrance conductors. They may be type TW, THW, RHW, or any type wire suitable for outdoor (wet) locations. They may be separate wires brought in through conduit and service head or wires made up into service-entrance cable approved for the purpose (Figure 11.4).

Service-entrance conductors are also governed by NEC.

1. NEC 230-40 (a)   Service-entrance conductors shall be insulated. Exceptions 1, 2, 3: Bare copper grounded conductor acceptable if in raceway or direct burial where suitable for soil conditions, or in soil when cable has moisture and fungus-resistant covering. Exception 4: Aluminum grounded conductor without individual insulation or covering acceptable if in raceway or direct burial when part of cable assembly having a moisture and fungus-resistant outer covering and approved for the purpose.
2. NEC 230-41 (a)   Service-entrance conductors shall have ampacity to carry loads as per Article 220.
3. NEC 230-41 (b)   Ungrounded conductors. 230-41 (b-1), (b-2) In single family dwellings with six or more 2-wire branch circuits or with initial computed load of 10 KW or more, conductors shall have ampacity of not less than 100 amperes 3-wires.
4. NEC 230-41 (b-3)   Not smaller than 60 amperes for other loads. Exception 1: Not smaller than No. 8 copper or No. 6 aluminum for

PARASYN® TW

60°C. BUILDING WIRE

UL FILE NUMBER E 14656

- insulation ....**PARASYN® 60 (PVC)**
- conductor ....**COPPER**

Paranite's PARASYN® type TW is an Underwriter's Laboratories listed type TW cable suitable for use at conductor temperatures up to 60°C in either wet or dry locations. The PARASYN® 60 (PVC) insulation has been specifically designed for high insulation resistance and outstanding moisture stability up to 60°C temperatures.

In addition to Underwriter's Laboratories type TW, PARASYN® TW exceeds the requirements of IPCEA-S-61-402 (3.7), ASTM-D-2219, U.L. 60°C in oil service and Federal Specification JC-30A. The solid bare copper conductors are made to ASTM-B-3 and stranded bare copper conductors to ASTM-B-8 and B-3.

Paranite's PARASYN® TW is available in all standard NEMA colors and each finished package is shipped with U.L. Insulated Wire Label attached.

Size	Number of Strands	Insulation Thick.		Approx. OD Inches	Approx. Weight Per 1000 Ft.	Ampacity**	Package*
		64th	Mils				
14	1	2	.030	.13	20	15	500 Ft. Ctn.
12	1	2	.030	.15	28	20	500 Ft. Ctn.
10	1	2	.030	.17	42	30	500 Ft. Ctn.
14	7	2	.030	.14	20	15	500 Ft. Ctn.
12	7	2	.030	.16	29	20	500 Ft. Ctn.
10	7	2	.030	.18	43	30	500 Ft. Ctn.
8	7	3	.045	.24	72	40	500 Ft. Ctn.

* 14-10 also available in 2500 Ft. Reels.
  8 also available in 1000 Ft. Reels.
** Ampacity based on NOT more than three conductors in raceway or conduit with an ambient temperature of (30°C) (86°F) and maximum conductor temperature of 60°C.

(a)

Figure 11.4 Service entrance conductors. (a) Type TW solid building wire. (Courtesy of Essex International)

more than two-2 wire branch circuits. Exception 2: Not smaller than No. 8 copper or No. 6 aluminum if by special permission for loads limited by demand or by source of supply. Exception 3: Not smaller than No. 12 copper or No. 10 aluminum for limited loads of a single-branch circuit but never smaller than the branch-circuit conductors.

5. NEC 230-54   Raintight service head required.
6. NEC 230-54 (c)   Service heads located above service drop conductors. Exception: where impracticable to locate service head above drops, it may be located not more than 24 inches to one side.

PARASYN® THHN - THWN
60°-105° BUILDING WIRE

UL FILE NUMBER E 14656 OR E 53446

jacket .... **NYLON**
insulation .... **PARASYN® 90 (PVC)**
conductor .... **BARE COPPER**

Paranite's Type THHN or THWN is a small diameter general purpose 600 volt building wire for use as power, lighting and control wiring. The thin, high grade PVC insulation and smooth, tough Nylon jacket permits the easy use of more wires in a given size conduit. The wide range of listings by Underwriter's Laboratories Standard 83 include: THWN - 75°C wet or dry, THHN - 90°C dry, machine tool wire 90°C*, appliance wire 105°C, and sizes 14 thru 500 MCM are gasoline and oil resistant 75°C.

The copper conductors meet the requirements of ASTM-B-3 if solid, and ASTM-B-3 and B-8 if stranded. As Type THHN or THWN, this wire also complies with Federal specification JC-30A.

Size	Number of Strands	Thickness In Inches		Approx. O.D. Inches	Ampacity**		Approx. Weight Per 1000 Ft.	Package***
		Insul.	Jacket		THHN 90°	THWN 75°C		
14	Solid	.015	.004	.11	15	15	16	500 Ft. Ctn.
12	Solid	.015	.004	.13	20	20	25	500 Ft. Ctn.
10	Solid	.020	.004	.16	30	30	38	500 Ft. Ctn.
14	19	.015	.004	.12	15	15	17	500 Ft. Spool
12	19	.015	.004	.14	20	20	25	500 Ft. Spool
10	7	.020	.004	.17	30	30	40	500 Ft. Spool
8	7	.030	.005	.22	50	45	65	500 Ft. Spool
6	7	.030	.005	.26	70	65	97	1000 Ft. Reel
4	19	.040	.006	.33	90	85	155	1000 Ft. Reel
3	19	.040	.006	.36	105	100	191	1000 Ft. Reel
2	19	.040	.006	.39	120	115	236	1000 Ft. Reel
1	19	.050	.007	.45	140	130	304	1000 Ft. Reel
1/0	19	.050	.007	.49	155	150	376	1000 Ft. Reel
2/0	19	.050	.007	.54	185	175	468	1000 Ft. Reel
3/0	19	.050	.007	.59	210	200	582	1000 Ft. Reel
4/0	19	.050	.007	.65	235	230	725	1000 Ft. Reel
250 MCM	37	.060	.008	.70	290	255	863	1000 Ft. Reel
300 MCM	37	.060	.008	.75	300	285	1024	1000 Ft. Reel
350 MCM	37	.060	.008	.80	325	310	1188	1000 Ft. Reel
400 MCM	37	.060	.008	.85	360	335	1350	1000 Ft. Reel
500 MCM	37	.060	.008	.93	405	380	1670	1000 Ft. Reel

* Solid and 7 Str. not marked or labeled as machine tool wire.

** Based on 3 conductors in conduit, with ambient temperature of (30°C) (86°F)

*** 14, 12 and 10 also available in 2500 Ft. Reels.

(b)

Figure 11.4 (b) Type THW stranded. (Courtesy of Essex International)

7. NEC 230-54 (e)   Service conductors brought out through separately bushed holes in service heads.
8. NEC 230-43   Service-entrance conductors may be installed as (1) open wiring on insulators; (2) rigid metal conduit; (3) elelctrical metallic tubing; (4) service-entrance cables; (5) wireways; (6) busways; (7) auxiliary gutters; (8) rigid nonmetallic conduit; (9) cablebus; or (10) mineral-insulated metal-sheathed cable.
9. NEC 300-18 (a)   Raceways to be installed as complete system before wires are drawn in. Figure 11.5.
10. NEC 230-26   In no case shall the point of attachment to the service drop be less than 10 feet above the finished grade.

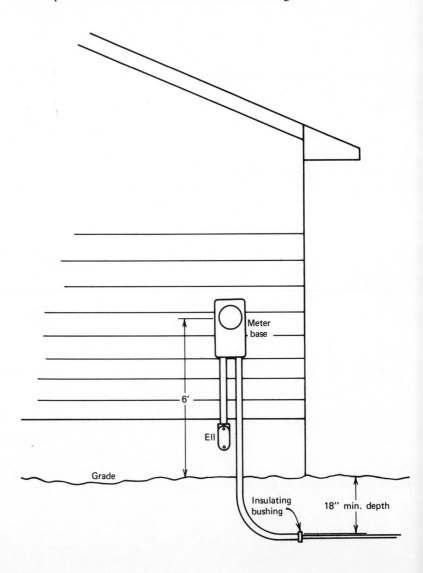

Figure 11.5 Underground service.

11-4 (Figure 11.5)

Service Laterals In most new residential developments and subdivisions, the choice is made to bring the service into the house underground. When the power supplier's conductors to a building are installed underground, they constitute a service lateral.

The term "service lateral"includes the underground cable installed from the point of connection to the utility company's system. The conductors usually terminate at the base of a pad-mounted transformer or in some instances a pedestal that is placed at or near the rear of the owner's property line, or it could be located in some other inconspicuous location in the development.

The service lateral conductors are usually furnished and installed at the owner's expense.

The underground service installation is covered by the National Electrical Code requirements.

1. NEC 230-30   Service lateral conductors shall be insulated. Exceptions 1, 2, 3: Bare copper grounded conductors acceptable if in raceway, or direct burial where suitable for soil conditioning, or in any soil when cable has moisture and fungus-resistant covering. Exception 4: Bare aluminum grounding conductor acceptable if in raceway, or direct burial if in a cable with moisture and fungus-resistant covering.
2. NEC 230-31   Conductors shall have sufficient ampacity to carry the load. Service laterals shall not be smaller than No. 8 copper or No. 6 aluminum, except for limited loads. Not smaller than No. 12 copper or No. 10 aluminum or limited loads.

When the service lateral conductors are installed by the electrician, the installation must conform to NEC 230-48 and 230-49, which refer to the protection of the conductors against damage and the sealing of underground conduits where they enter the house. NEC 230-49 refers to 300-5, which covers all situations relative to underground wiring.

11-5

The Meter Socket The utility company decides on the location of the meter. In most cases it will be the outside type, with its socket exposed to the weather. This location is helpful to the meter reader, since access to the inside of the house is not needed.

The meter socket is sometimes furnished by the utility company and sometimes by the owner. In any event it is installed by the electrician. The electrician must fasten the meter socket to the building, install the service conduit and wires, and then cut in conductors and attach to the clips in the meter socket. The supply conductors are connected to the upper terminals and load conductors to the bottom.

The neutral conductor may be grounded in the meter socket and in the service panel. However, local utility company requirements prevail.

In some localities, a combination of meter socket, service disconnect switch, and circuit breakers in the same enclosure is favored (Figure 11.6).

The recommended height of a meter socket is between 5 feet and 6 feet, at eye level.

(a)

Figure 11.6 (a) Combination rainproof underground service, surface-mounted meter base, service disconnect switch-circuit breakers, (b) Combination rainproof service entrance with branch circuit breakers. (Courtesy of Square D Company)

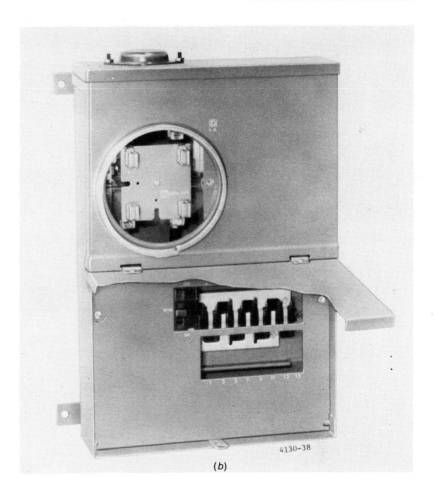

(b)

11-6 Service Distribution Panel All of the circuits you have studied so far are known as branch circuits. Chapter 4 instructed us, "a branch circuit is the circuit conductors between the final overcurrent device protecting the circuit and the outlets."

Types of Circuits 1. A general-purpose branch circuit is a circuit that supplies a number of outlets for lighting and small appliances.

2. An appliance branch circuit is a circuit that supplies energy to one or more outlets to which kitchen appliances are connected; no lighting fixtures are permitted on this circuit.

3. An individual branch circuit is a circuit that supplies power to only one utilization equipment, such as a kitchen electric range.

(c)

1140-18

Figure 11.6 (c) Combination rainproof surface-mounted main circuit breaker and service entrance, with threaded conduit hubs on top, knockouts available for bottom line and load. (Courtesy of Square D Company)

In Chapter 5, review the drawings of the branch circuit cable runs. The diagrams indicated a cable with an arrow point and number. This represented the number of the circuit and cable home run to the main power distribution panel.

Types of Service Panels There are many types of service distribution panels used in residential wiring; however, there are only two types of overcurrent protection: the fuse and the circuit breaker.

The circuit breaker panel is usually installed with a main disconnect circuit breaker and facilities to mount plug-in circuit breakers (Figure 11.7). The circuit breaker is a device that combines the function of an overcurrent circuit protective device with that of a switch. Basically it is a switch equipped with a tripping mechanism that is activated by excessive current (Figure 11.8). The important fact to remember is that circuit breakers, like fuses, react to excess current. However, the breaker is not self-destructive. After the breaker opens to clear the fault, it can simply be reclosed and reset. This means that replacement problems are eliminated.

To install a circuit in a panel, the slot at the rear of the breaker is engaged over the protruding metal stud in the panel. The breaker jaws are firmly pushed onto the bus-bar tab. Breakers must be clearly marked to show whether they are in the open (OFF) or closed (ON) position. Trip the breaker to the OFF position and you are ready to connect the branch circuit wire. Use a 15-ampere breaker for No. 14 AWG wire, 20 ampere breaker for No. 12 wire, 30 ampere for No. 10 wire, and so on.

Circuit breakers are made in single-pole and double-pole with voltage and current ratings to match the circuit conductors.

In modern installations, the devices to control and protect individual branch circuits are always included in the same distribution panel cabinet and must bear the UL label and must be marked as suitable for service equipment.

Fuse panels for main service distribution panels usually consist of a main pull-out fuse, range pull-out fuse, and other branch circuit plug fuse holders (Figure 11.9). When installing a new fuse panel, the NEC states that only "Type S fuses must be used." Type S fuses have a different base size for identifying the current range of the fuse; 15-ampere, 20-ampere, and 30-ampere. Type S fuses are so designed that they cannot be used in any fuse box other than a type S fuseholder or a fuseholder with a Type S adapter inserted (Figure 11.10).

It is the reponsibility of the electrician to determine the ampere rating of the circuit and select the proper size Type S adapter. The adapter is then inserted into the Edison-base fuseholder. The

(a)

Figure 11.7 Main disconect panels. (*a*) Service disconnect panel showing main disconnect switch, cover and fill of safety-breakers. (Courtesy of Cutler-Hammb, Incorporated) (*b*) Flush type load center with main circuit breaker and neutral terminals. Complete panel before installation of trim. (Courtesy of General Electric Company—Circuit Protective Devices Department)

(*b*)

adapter makes the fuseholder nontamperable and noninterchangeable. The proper size fuse can now be inserted into the adapter, and the wire for that circuit may be connected to the fuseholder terminal.

Some residential equipment such as counter-mounted cooking tops, ovens, ranges, clothes dryers, water heaters, and electric

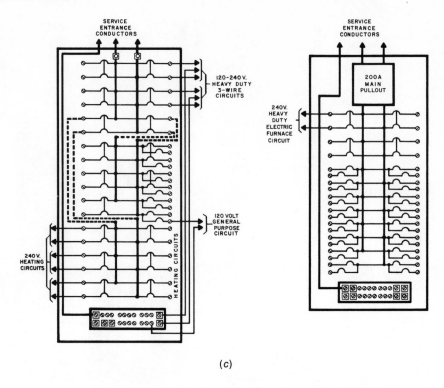

(c)

Figure 11.7 (c) Internal wiring diagrams for breaker type service equipment.

space heating operate on 240 volts. This requires a different type of fuse: the ferrule contact cartridge type. A ferrule cartridge fuse is made of two different physical sizes. The smaller of the two is produced in current ratings up to 30 ampere. The larger is available in 35- to 60-ampere ratings and is made in 5-ampere units (35, 40, 45, 50, 60). (Figure 11.11).

When currents in excess of 60 amperes are flowing in a circuit (usually the main fuses), a more rugged knife-blade cartridge type fuse is used to protect the entrance conductors.

The ground-fault circuit interrupter receptacle is covered in Chapter 4, page 54. However, in this section the circuit breaker that protects both the equipment and people will be introduced for discussing. It is a circuit breaker with a built-in ground-fault circuit interrupter. It provides the same branch circuit overcurrent protetion as a standard circuit breaker; however, it provides added protection from ground-faults. It is designed and manufactured to automatically open a circuit when a fault current is 5 milliamperes or more.

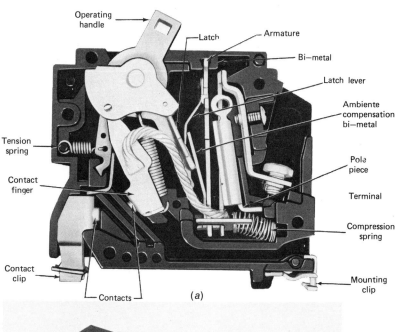

Operating handle

Latch

Armature

Bi—metal

Latch lever

Ambiente compensation bi—metal

Pole piece

Terminal

Compression spring

Tension spring

Contact finger

Contact clip

Contacts

(a)

Mounting clip

15

(b)

Figure 11.8 (a) Typical circuit breaker (open view). (b) Single pole breaker. (Courtesy of Cutler-Hammb, Incorporated)

(c)

Figure 11.8 (c) Double pole 15-ampere breaker. (Courtesy of Cutler-Hammer Incorporated)

The GFCI breaker is a self-contained unit that may fit directly into the distribution panel (Figure 11.12). It operates on the principle that the current leaving a circuit is equal to the current entering that circuit. Both supply conductors of the circuit pass through a highly developed transformer. When there is no leakage current, the magnetic fields around the supply conductors cancel one another. No voltage is produced in the transformer.

Should a leak or fault develop, more current is entering the circuit on one supply conductor that is leaving on the other. This magnetic imbalance causes a voltage to be induced into the transformer coils. An amplifier increases the strength of the voltage and uses it to trip the circuit breaker. A fault current as low as 0.002 amperes will trip the GFCI breaker.

GFCI circuit breakers or a GFCI receptacle are required in the following areas:

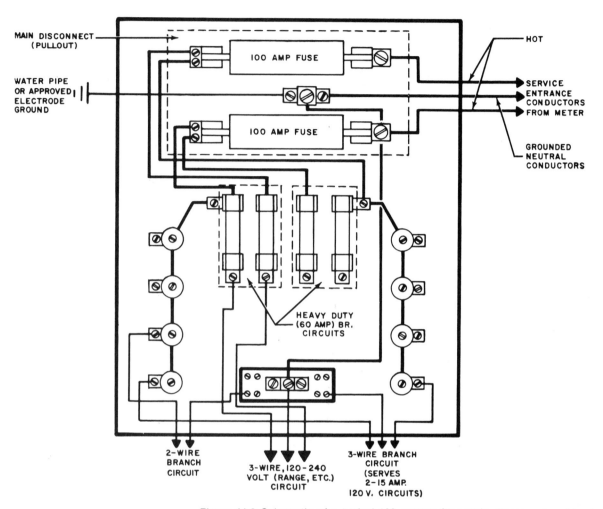

Figure 11.9 Schematic of a typical 100-ampere fuse main disconnect and load center.

NEC 210-8 In residential occupancies, on all 120-volt single phase, 15-ampere, 20-ampere receptacle outlets outdoors, in garages, and in bathrooms.

NEC 210-8 On the construction site, (usually the temporary power pole).

NEC 680-6 Swimming Pools: lighting fixtures and lighting outlets.

NEC 680-20 Underwater lighting fixtures (wet-niche).

NEC 680-31 Storable swimming pools.

NEC 680-41 (a) Supplying fountain equipment.

 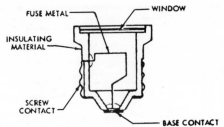

(a)

Left: Plug fuse designed to pass safely 15 amps. Note fuse link in hexagonal shaped window. Right: Cross section of plug fuse. (Bussman Mfg. Co.)

Adapter to accept special fuse sizes. Barb on lower front prevents easy removal of adapter. (Bussman Mfg. Co.)

(b)

Type "S" fustat for use with special adapter and certain amperage ratings. (Bussman Mfg. Co.)

(c)

Figure 11.10 Fuses. (a) Plug fuse. (b) Type "S" fustat. (c) Adapter to accept special fuse sizes. (Bussman Manufacturing Company)

NEC 680-46 Fountains: all power supply cord-plug connected equipment.

NEC 680-5 Swimming pool transformers.

NEC 555-3 Marinas and boatyards.

NEC 517-13 Health care facilities.

NEC 215-9 Feeders.

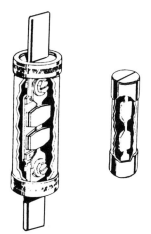

Cutaway of fuse types available in renewable and non-renewable links. Left: Blade-contact fuse available in sizes larger than 60 amps. Right: Furrule-contact cartridge fuse available in sizes to 60 amps. Both fuses shown are renewable type. Fuse links may be replaced. (Bussman Mfg. Co.)

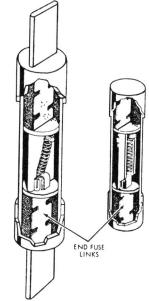

END FUSE LINKS

Cutaway view of dual element fuses (non-renewable). Both sizes are of the time-delay type. (Bussman Mfg. Co.)

Figure 11.11 Cartridge fuses. (Bussman Manufacturing Company)

11-7 Grounding The purpose of grounding is safety. According to the NEC, a wiring system must afford protection to life and property against faults caused by electrical disturbances, lightning, failure of electrical equipment that is a part of the wiring system, or failure of equipment and appliances that are connected to the wiring system.

For this reason, all metal enclosures of the wiring system, as well as the noncurrent-carrying or neutral conductors, should be tied together and reduced to a common earth potential.

Grounding falls into two categories: (1) system grounding (2) equipment grounding.

1. *System Grounding:* System grounding means the connection of the neutral conductor of the wiring system to the earth. Its purpose is to drain off any excessive high voltage that may accidentally enter the system as a result of an electrical disturbance, lightning, an insulation breakdown in the supply transformer, or an accidental contact between the service conductors and high tension wires. An accidental grounding of one of the current carrying conductors will result in a short circuit causing

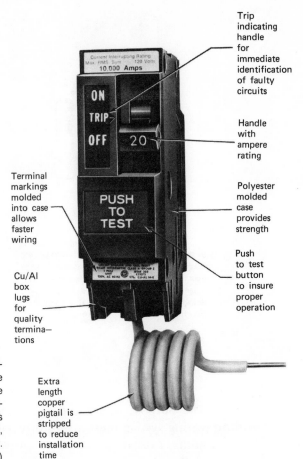

Current Interrupting Rating
Max. RMS Sym.   120 Volts
10,000 Amps

ON
TRIP
OFF
20

**PUSH
TO
TEST**

Trip
indicating
handle
for
immediate
identification
of faulty
circuits

Handle
with
ampere
rating

Polyester
molded
case
provides
strength

Push
to test
button
to insure
proper
operation

Terminal
markings
molded
into case
allows
faster
wiring

Cu/Al
box
lugs
for
quality
termina-
tions

Extra
length
copper
pigtail is
stripped
to reduce
installation
time

Figure 11.12 Ground fault circuit interruption feature is built into a single pole molded case circuit breaker of the type usually used to protect normal receptacle circuits. These circuit breaker units will fit into an ordinary load center, replacing a conventional circuit breaker. (Courtesy of General Electric Company)

a fuse or circuit breaker to open, thereby disconnecting the line conductors.

2. *Equipment Grounding:* Equipment grounding, or grounding noncurrent-carrying parts of the installation, means the steel raceway itself, service equipment panel, and metal enclosures of equipment, like the frames of ranges or motors.

The purpose of this grounding is to prevent a voltage higher than earth potential on the enclosures or equipment, reducing the danger of shock in case a life-conductor comes in contact with these conductive parts.

The grounding is accomplished by running a wire from the neutral connection in the main service switch or, in some cases, the meter socket, to the water piping system on the street side of the water meter.

The NEC again rules here:

NEC 250-81 Where available on the premises, a metal underground water pipe shall always be the grounding electrode.

NEC 250-92 (a) Metal grounding electrode conductor enclosures shall be electrically continuous from cabinet to grounding electrode.

NEC 250-92 (a) If metal enclosure is not electrically continuous it must be bonded at each end to grounding conductor.

NEC 250-115 Ground clamp approved for the purpose shall be used to connect grounding electrode conductor to grounding electrode.

NEC 250-112 Connection should be on street side of water meter or bonding required around valves, meter, unions, etc.

Size of grounding wire is found in NEC Table 250-94.

11-8 All service equipment must be properly bonded.

Bonding The purpose of installing bonding bushings and bonding jumpers on service-entrance equipment is to assure a low impedance path to ground should a fault occur on any of the service entrance conductors.

1. NEC — Definitions: Main Bonding Jumper is the connection between grounded circuit conductor and equipment grounding conductor at service.
2. NEC 250-79 (a) (b) Equipment Bonding Jumper — main and equipment bonding jumper shall be (a) copper or other corrosion-resistant metal: (b) attached as per NEC 250-113 and 250-115.
3. NEC 250-79 (c) Equipment bonding jumper on supply side of service and main bonding jumper. Sized as per Table 250-94 but not less than 12½ percent of largest phase conductor.

Summary
1. All service equipment wiring must adhere to the National Electrical Code.
2. The local utility company is responsible for locating the meter and service drop.
3. The meter socket is grounded for protection.
4. Grounded neutral conductors are not switched or protected by an overcurrent device.
5. Circuit breakers and fuses are safety devices to protect wire against short circuits and overloads.

6. The meter is located outside of the house for convenience.
7. Service laterals are protected by the NEC.
8. Circuit breaker panels are widely used in residential wiring.
9. To reset the circuit breaker, move handle to reset position.
10. The purpose of grounding a service is for safety.

Problems 11-1 When circuit breakers trip due to a short circuit or overload, how is power restored? Explain.

11-2 What part of a circuit breaker causes the breaker to trip:
  a. on an overload?
  b. on a short circuit?

11-3 A branch circuit wired with No. 12 wire would be protected with a fuse rated at:
  a. 30 amperes
  b. 15 amperes
  c. 20 amperes

11-4 A branch circuit is the circuit conductors between the final over-current device protecting the circuit and the outlets. It is known as an appliance branch circuit when it supplies:
  a. a single receptacle outlet for one large appliance only.
  b. a number of outlets for lighting and all appliances.
  c. a number of outlets for appliances only.

11-5 List six parts of the electrical service.

11-6 Explain the function of a ground-fault circuit interrupter breaker.

Chapter 12 **Service Load Calculations**

Instructional Objectives
1. To learn how to calculate the size of the service entrance conductor.
2. To learn how to use the National Electrical Code in calculating service load calculations.
3. To determine by the National Electrical Code when a neutral conductor may be reduced in size.
4. To provide the basic information to calculate a dwelling.
5. To become more familiar with the NEC.

Self-Evaluation Questions

Test your prior knowledge of the information of this chapter by answering the following questions. Watch for the answers as you study the chapter. Your final evaluation of whether you understand the material is measured by your ability to answer these questions. When you have finished the study of this chapter, return to this section and answer the questions again.

1. What is the unit load per square foot of floor space for the general lighting load of a dwelling?
2. What demand factors are allowed when computing a general lighting load in a dwelling?
3. Is the above demand factor used when computing the lighting load in residential occupancies?
4. Why is it important to calculate the service load?
5. Is the size of the neutral service conductor ever reduced?
6. What will result if the service conductors are too small?

12-1 Importance of Service Load Calculations

Adequate service capacity is an absolute necessity. When calculating and installing electric space heating, the loads are sized to operate at full capacity, during design temperature conditions, for relatively long periods of time. NEC 220-15. If an undersized electrical service is installed, full capacity operation will reduce the power available to the household appliances such as the electronic range and oven or counter-mounted cooking top, dishwasher, trash compactor, hot water heater or the television set. If overloading the service becomes severe enough it can cause conductor heating and contribute to premature insulation breakdown and thus cause a fire hazard. To avoid overloading the electrical service, especially where electrical space heating equipment is to be installed, service load calculations are mandatory. Most requests for electrical inspection permits require them. Review Figure 1.4.

12-2
Standards for Service
Load Calculations

The National Electrical Code: In determining the lighting load on the watts per square foot basis, the floor area shall be computed from the outside dimensions of the dwelling. Not included in this area are: open porches, garages, and unfinished or unused spaces (unless these spaces are adaptable for future use; then they shall be computed in the floor dimensions). See Chapter 3, Section 3-1.

NEC Table 220-2 (b) shows 3 watts per square foot.

NEC Section 220-16 (a) rates the two small appliance circuits at 3000 watts (1500 watts each circuit).

NEC Table 220-19, Column C, shows the load for two cooking units to be 65 percent of the sum of the nameplate rating of the two units.

NEC Section 220-16 (b) rates the laundry circuit at 1500 watts.

12-3
Calculate the Minimum
Service Load

With the above code sections and tables we are able to calculate the minimum service load for a dwelling 40 feet by 25 feet, or 1000 square feet. (Figure 12.1.)

Step 1 Calculate the lighting load requirements.

1000 sq ft × 3 watts per ft = 3000 watts.
(2 circuits for lighting load)

Step 2 Calculate the small appliance circuits code states.

We use two circuits = 3000 watts.
So far in our calculations we have 4 circuits and 6000 watts.

Step 3 It is reasonable to believe that not all lights and outlets for

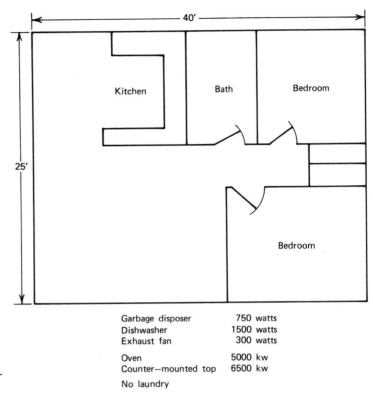

Figure 12.1 Typical single-family dwelling floor plan.

Garbage disposer	750 watts
Dishwasher	1500 watts
Exhaust fan	300 watts
Oven	5000 kw
Counter—mounted top	6500 kw
No laundry	

the appliance circuits will be in use at the same time, so a demand factor is used to calculate the feeder demand factor.

NEC Table 220-11 shows the first 3000 watts of lighting and appliance circuits shall be computed at 100 percent. The remainder of the 3000 watts are to be computed at 35 percent.

Example:

In step 1 we showed a total of 6000 watts.

First 3000 watts computed at 100%	=	3000 watts
Remainder, 3000 watts computed at 35%	=	750 watts
Our 6000 watts are now reduced to		3750 watts

Step 4 Feeder load for the oven and countertop.

Oven is rated at	5000 watts
Countertop rated at	6500 watts
	11500 watts

NEC Table 220-19 Column C shows that we can take a demand factor of 65 percent of the sum of the nameplate rating of the two units.

Example:
Again we use the demand factor—65%
65% of 11,500 watts = 7475 watts.
Our 11,500 watts now reduced to 7475 watts.

Step 5 Compute the feeder load.

1. Lighting and appliance circuits	3750 watts
2. Countertop and oven	7475 watts
3. Dishwasher (cannot be reduced)	1500 watts
4. Garbage disposer unit	750 watts
5. Exhaust fan	300 watts
	13,775 watts

Step 6 Compute the feeder size.

The utility company has supplied a 115/230 volt overhead service to the house.

The total computed feeder load is 13,755 watts.

Example:
13,755 watts ÷ 230 volts = 59.8 amperes.

Table 310-16 of the NEC shows that 59.8 amperes require a #6 AWG Type THW copper conductor good for 65 amperes.

Table 3A, Chapter 9 of the NEC shows that a 1 inch conduit is required for 3 #6 AWG, Type THW, copper conductors.

12-4 Calculate Service Load for 1500 sq ft Dwelling (Figure 12.2).

General lighting load 1500 sq. ft. × 3 =	4500 watts
4500 ÷ 115 volts = 39.1 amperes (3-15 amp circuits)	
Allowance for two appliance circuits	3000 watts
(2 circuits—20 amps)	
Laundry circuit NEC 220-3(a)	1500 watts
(1 circuit—20 amps)	
NET TOTAL	9000 watts
Feeder demand factors	
First 3000 watts at 100%	= 3000 watts
Remainder 6000 watts at 35%	= 2100 watts
NET TOTAL	5100 watts
12 kW Range Load (Table 220-19)	8000 watts
NET LOAD WITH RANGE	13,100 watts
Utility installed 115/230 volt 3 wire service.	

Example:
13,100 watts ÷ 230 volts = 57 amperes.

NEC 230-41(b) Net computed load exceeds 10 kW, so service conductors shall be at 100 amperes.

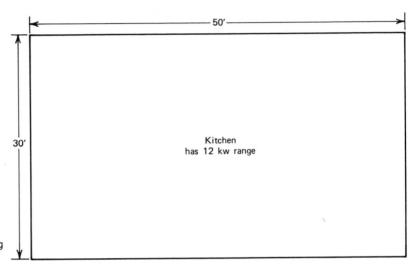

Figure 12.2 Floor plan for calculating service load.

12-5 Reducing the Feeder Service Neutral Many electricians prefer to reduce the service neutral conductor. However, for most installations and especially for a dwelling equipped with an electric range, electric clothes dryer, and electric hot water heater, a neutral conductor one size smaller than the ungrounded conductors is usually more than adequate. It is recommended that the correct neutral size first be determined using the procedure outlined in this chapter.

NEC 220-22 will allow us to reduce the service neutral.

Example:
Using the calculated service load for 1500 sq ft dwelling in 12.4.
Lighting and appliance load 5100 watts
Range load reduced 70% 8000 × 70% = 5600 watts
10,700 watts
10,700 watts ÷ 230 volts = 46.5 amperes.

Summary
1. When calculating electric space heating equipment, loads are figured to operate at full capacity.
2. Severe overloading the service could cause conductor heating and insulation breakdown.
3. Floor area is computed from the outside dimensions of a dwelling.
4. Small appliance circuits are rated at 3000 watts.
5. Demand factors are used when calculating service feeders.
6. The National Electrical Code has many requirements relative to service load calculations.

Problems
12-1 Calculate the lighting load for a 1575 sq ft dwelling. Show calculations and code sections.
12-2 Compute the lighting, two appliance and a laundry circuit, for a new 1625 sq ft dwelling. Show calculations and code sections.
12-3 What is the load for an electric range rated at 16 kW?
12-4 When the service entrance calculation results in a computation of 10 kW or more, what is the minimum size service permitted by the NEC?
12-5 Calculate the service load for a 1550 sq ft dwelling. The kitchen has a 750 watt garbage disposer, 1200 watt dishwasher, and 350 watt exhaust fan. Show calculations.
12-6 Calculate THW copper conductor and conduit size for question 12-5 above. Do not reduce neutral.
12-7 What load rating may be used for an electric range rated at not over 12 kW?
12-8 How many circuits will be needed for question 12-5?

References

National Fire Protection Association, *National Electrical Code.* New York: McGraw-Hill, 1977.

W. J. McGuinness and B. Stein, *Building Technology: Mechanical and Electrical Systems,* New York: Wiley, 1977.

H. P. Richter, *Practical Electrical Wiring*, Tenth Edition, New York: McGraw-Hill, 1976.

R. C. Mullin, *Electrical Wiring Residential,* New York: Delmar, 1975.

W. N. Alerich, Electrical Construction Wiring. American Technical Society, Chicago, 1971.

Sears, Roebuck and Co. "Simplified Electrical Wiring," 1969.

American Heating and Air Conditioning Wholesalers Association, *Fundamentals of Heating,* Second Edition. Columbus, 1971.

Energy Utilization Systems, "Energy Consumption and Life-Cycle Costs of Space Conditioning Systems." Pittsburgh. 1976.

ALA Research Corporation, "Solar Dwelling Design Concepts." Washington, D.C., 1976.

Glossary

AWG. American Wire Gauge, the standard wire size measuring system in the United States.

Air Changes. The number of times the air is changed per hour in a room.

Air Conditioning. A method of filtering air and regulating its humidity and temperature in buildings.

Air Density. The weight of air, pounds per cubic foot.

Ampacity. A wire's ability to carry current safely, without undue heating. The term formerly used to describe this characteristic was current-carrying-capacity of the wire.

Appliance Circuit. A branch circuit that supplies outlets specifically intended for appliances.

Appliance Outlet. An outlet connected to an appliance circuit. It may be a single or duplex receptacle or an outlet box intended for direct connection to an appliance.

Architectural-Electrical Plan. Architectural plan on which electrical work is shown.

Auxiliary Resistance Heating. Electric resistance heaters that supplement the heat from a heat pump or solar unit.

Baseboard. An electric resistance heater along the baseboard.

Bonding. To assume a low impedance path to ground should a fault occur on any of the service entrance conductors.

Branch Circuit. Wiring between the last overcurrent device and the branch circuit outlets.

British Thermal Unit. Quantity of heat.

Cable. An assembly of two or more wires or a single wire larger than No. 8 AWG.

Centralized. A system with one heating or cooling source and a ductal distribution network.

Circuit. An electrical arrangement requiring a source of voltage, a closed loop of wiring, an electric load, and some means for opening and closing it.

Circuit Breaker. A switch-type mechanism that opens automatically when it senses an overload (excess current).

Circular Mil. Unit of measurement for determining the area of a wire in cross section, equal to the area of a circle having a diameter of 1 mil. (0.001 in.).

Clearing A Fault. Eliminating a fault condition by some means. Generally taken to mean operation of the overcurrent device that opens the circuit and clears the fault.

Common Neutral. A neutral conductor that is common to, or serves, more than one circuit.

Connected Load. The sum of all loads on a circuit.

Convector. A heating element that warms the air passing over it which, in turn, rises to warm the space by convection.

Convenience Outlet. A duplex receptacle connected to a general purpose branch circuit, not intended for any specific item of electrical equipment.

Current. The electric flow in an electric circuit, expressed in amperes (amps).

Degree Day. The number of Fahrenheit degrees that the average outdoor temperature over a 24hour period is less than 65° F.

Demand Factor. Ratio of the maximum amount of power required by a system to the total connected load of the system.

Distribution Panel. A control center that distributes electrical energy to branch circuits.

Electronic Air Cleaner. A filter somewhat more efficient than the usual filter, particularly for the removal of small suspended particles.

Entrance Cable. Fiber-insulated cable of two large-sizee insulated wires covered with a bare spiral wound sheathed that serves as the neutral conductor.

Entrance Head. A watherproof housing for supporting the entrance cable.

Fault. A short circuit.

Fossil Fuels. Oil, gas, and coal.

Four-Way Switch. A four-terminal switch used to control outlets from three or more locations.

Furnace. A unit that warms the air in warm air heating systems.

Fuse. A protective device in which excessive current meets the fuse element and opens the circuit.

GFCI. Ground fault circuit interrupter—a device that senses ground faults and reacts by opening the circuit.

Gang. One wiring device position in a box.

Ganged Switches. A group of switches arranged next to each other in ganged outlet boxes.

General Purpose Branch Circuit. One that supplies a number of outlets for general lighting and convenience receptacles.

Ground. Any point connected to ground, may be metallic frame or common bus.

Ground Bus. A busbar in a panel, deliberately connected to ground.

Ground Conductor. A conductor run in an electrical system, which is deliberately connected to the ground electrode. Used to provide a ground point throughout the system. A green-colored insulated conductor.

Ground Electrode. A piece of metal physically connected to ground. Can be pipe, rod, mat, pad, or structural member.

Ground Fault. An unintentional connection to ground.

Grounded. Connected to earth or a conducting body.

Heat Pump. An all-electric heating/cooling device that takes energy for heating from outdoor air.

Home-Run. Cables that run between the service distribution panel and the first outlet in the branch circuit.

Hot Wire. The black or red insulated conductor that is electrically energized.

Individual Branch Circuit. One that supplies only a single piece of electrical equipment.

Infiltration. Cold air that leaks in. Expressed as air changes per hour.

Junction Box. Metal or non-metallic box in which tap to circuit is made.

Labeled. Materials and equipment having a label indicates compliance with standards or tests to determine suitable usage.

Line Voltage Thermostat. A thermostat that is connected directly to the line. Full line voltage is fed through it to the controlled heater.

MCM. Thousand circular mil—used to describe large wire sizes.

Meter Base. A device intended to hold the kilowatt-hour meter.

Multipole. Connects to more than 1 pole, such as a 2-pole circuit breaker.

Neutral Conductor. The circuit conductor with white or gray insulation that is normally grounded. It should never be fused or disconnected.

Nonmetallic Sheathed Cable. Single or multiple conductors within a fibrous protective covering.

Overcurrent Device. A fuse or circuit breaker designed to protect a circuit against excessive current by opening the circuit.

Overload. A condition of excess current; that is more current flowing than the circuit was designed to carry.

Panelboard. A box containing a group of overcurrent devices intended to supply branch circuits.

Panel Directory. A listing of the panel circuits appearing on the panel door.

Panel Schedule. A schedule appearing on the electrical drawing detailing the equipment contained in the panel.

Raceway. Any support system for the containment and protection of electric wires.

Radiant Cables. Electric cables embedded in the ceiling for heating.

Range Hood. Hood over a stove to collect odor-laden air that is to be exhausted.

Receptacle Wires. Number of connecting wires, including the ground wire.

Romex. One of several trade names for type NM nonmetallic-sheathed cable.

R-Value. The resistance rating of thermal insulation

Service Drop. The overhead service wires that serve a dwelling.

Service Lateral. The underground service conductors.

Service Switch. A disconnect switch or circuit breaker. Purpose is to completely disconnect the dwelling from the electric service.

Short Circuit. An electrical fault.

Single-Pole Switch. Capable of opening and closing one side of a circuit.

Stripping. To remove the insulation from a conductor.

Switch. A device used to open or close a circuit.

Thermal Transfer. Moving heat into or out of occupied space.

Three-Way Switch. A three-terminal switch that connects two switching locations.

Toggle-Type Switch. A device with a projecting lever whose movement causes contact to be made with a snap action.

Transducer. A device that will convert one form of energy into another form.

Transformer. A device consisting of one or more coils for introducing mutual coupling between electrical circuits.

Travelers. Hot black or red conductors in a 3-wire circuit controlled by a 3-way switch.

U Coefficient. The rate of heat transmission.

Ventilation. Controlled air.

Voltage. The electric pressure in an electric circuit, expressed in volts.

Wire-Nut. Trade name for small, solderless, twiston branch circuit conductor connector.

Wiring Diagram. A diagram showing actual wiring, with numbered terminals. All wiring is shown.

Wiring Device. Receptacle, switch, light, dimmer switch, or any device that is wired in a branch circuit and fits into an outlet box, usually 30 amperes or smaller.

Zone. Section of a heating and/or cooling system separately controlled.

Index